BIHAR AND ORISSA DISTRICT GAZETTEERS

MONGHYR

BIHAR AND ORISSA DISTRICT GAZETTEERS

Patna/ **L.S.S. O'Malley**
Shahabad/ **L.S.S. O'Malley**
Monghyr/ **L.S.S. O'Malley**
Saran/ **L.S.S. O'Malley**

BENGAL DISTRICT GAZETTEERS

Burdwan/ **J.C.K. Paterson**
Darjeeling/ **L.S.S. O'Malley**
Howrah/ **L.S.S. O'Malley**
Hooghly/ **L.S.S. O'Malley**
Santal Parganas/ **L.S.S. O'Malley**
Sambalpur/ **L.S.S. O'Malley**
Puri/ **L.S.S. O'Malley**
Gaya/ **L.S.S. O'Malley**

BIHAR AND ORISSA DISTRICT GAZETTEERS

MONGHYR

BY

L.S.S. O'MALLEY, I.C.S.

REVISED EDITION

LOGOS PRESS
New Delhi-110 002

ISBN : 81-7268-135-6

First Published 1926

Reprinted 2007

Printed and Published by

LOGOS PRESS

Building No. 4788-90, Street No. 23, Ansari Road, Darya Ganj,
New Delhi-110 002 (India)

Phone : 23272187

PREFACE TO ORIGINAL EDITION.

I DESIRE to express my obligations to Mr. H. F. Samman, I.C.S., Collector of Monghyr, and Mr. P. W. Murphy, I.C.S., Settlement Officer, Bihar, for their assistance in the compilation of this volume. Much valuable information has also been obtained from the Final Report on the Survey and Settlement of North Monghyr (1905—7), by Mr. H. Coupland, I.C.S.

L. S. S. O'M.

PREFACE TO REVISED EDITION.

I DESIRE to acknowledge to Mr. M. J. Dixon, I.C.S., Collector of Monghyr, to the Rev. Dr. J. M. Macphail, and to Rai Bahadur Charu Chandra Mukharji, and other local officers of Government and the District Board for the assistance which they have given me. Chapter XIII has been contributed by Mr. G. E. Fawcus, Director of Public Instruction; and the greater part of Chapter IV by Major A. N. Palit, I.M.S., Civil Surgeon of Monghyr.

J. F. W .J.

PLAN OF CONTENTS.

TABLE OF CONTENTS.

CHAPTER I.

PHYSICAL ASPECTS.

CHAPTER II.

HISTORY.

CHAPTER III.

THE PEOPLE.

CHAPTER IV.

PUBLIC HEALTH.

CHAPTER V.

AGRICULTURE.

CHAPTER VI.

NATURAL CALAMITIES.

CHAPTER VII.

RENTS, WAGES AND PRICES.

CHAPTER VIII.

OCCUPATIONS, MANUFACTURES AND TRADE.

CHAPTER IX.

MEANS OF COMMUNICATION.

CHAPTER X.

LAND REVENUE ADMINISTRATION.

CHAPTER XI.

GENERAL ADMINISTRATION.

CHAPTER XII.

LOCAL SELF-GOVERNMENT.

CHAPTER XIII.

EDUCATION.

CHAPTER XIV.

GAZETTEER.

APPENDICES

GAZETTEER

OF THE

MONGHYR DISTRICT.

CHAPTER I.

PHYSICAL ASPECTS.

GENERAL DESCRIPTION.

THE district of Monghyr, the most westerly district of the Bhagalpur Division, is situated between 24° 22′ and 25° 49′ north latitude and between 85° 36′ and 86° 51′ east longitude. It extends over an area of 3,968 square miles, and has a population, according to the census of 1921, of 2,029,965 persons. For administration purposes it is divided into three subdivisions—Monghyr, Jamui and Begusarai, the headquarters being at Monghyr on the southern bank of the river Ganges. Monghyr is a corruption of the vernacular name Munger, the derivation of which is the subject of several theories, which will be dealt with in the article on the town in Chapter XIV.

Boundaries.

The district is bounded on the north by the districts of Bhagalpur and Darbhanga, on the east by Bhagalpur, on the south by the Santal Parganas and Hazaribagh, and on the west by Gaya, Patna and Darbhanga. In some places the beds of different rivers and streams form natural boundaries, but for the most part the boundaries are artificial.

Configuration.

The Ganges flows through the district from west to east, dividing it into two portions of unequal size and of very different character. The northern and smaller portion is a flat alluvial plain traversed by the Burh Gandak river, which flows through it from north-west to south-east. The country to the west of that river is a continuation of the level, well cultivated plains of Tirhut and grows rich spring (*rabi*) and autumn (*bhadoi*) crops. The country to the east is intersected by the Tiljuga and Baghmati rivers, is seamed by deserted channels, and is low-lying, swampy, and liable to inundation during the rains. The south of the district is also to a great extent alluvial, but the general level is higher, the surface is more undulating, and a large area is composed of hills and valleys covered with forest trees or scrub jungle. The wide difference in the characteristics of the country to the north and south of the Ganges has been graphically described as follows by a former Collector of Monghyr, Mr. E. Lockwood, in *Natural History, Sport and Travel* :—

"The northern part is an extensive plain formed by the rich alluvial soil brought down by the ever-changing river, while the southern portion consists of vast rice tracts and forests, which cover the metamorphic hills extending far away into Central India from the town of Monghyr. Lovers of natural history who visit Monghyr, find that this division of the district separates also, in a very marked manner, the most conspicuous species of the animal and vegetable kingdoms; and the sportsman who to-day may find tigers, bears, baboons, tupaias, peacocks, jungle-fowl and grey partridges in the undulating country to the south, will look in vain for such things if to-morrow he crosses the river northward.

"The river separates also the most conspicuous trees and plants. In the forest of the south are found the ebony tree (*diospyros melanoxylon*), the sal (*shorea robusta*), the most useful of all Indian timber trees, and the *mahua* (*bassia latifolia*), which supplies not only food to the lower classes, but also spirit to the drinking portion of the population. The south also yields vast quantities of rice, and a hundred and fifty tons of opium, grown on twenty-five thousand acres of land, whilst, after crossing the Ganges, little rice and not a single poppy will be seen.* In the north nine-tenths of the trees are cultivated mangoes, whilst wheat, Indian corn, various kinds of millet, peas, *masur* (*cicer lens*), *rahar* (*cytisus cajar*), oats, indigo, mustard, linseed and castor oil, are the principal crops which the landholders find profitable to grow. The northern portion of the district also, during the cold weather, forms a vast feeding ground for swimming and wading birds. On some of the marshes a hundred

* The area under poppy cultivation decreased in 1907-8 to 15,000 acres yielding 54 tons of opium. The cultivation has now been abandoned.

thousand ducks may be seen, so close together that they almost hide the water; and as flock after flock pass overhead on being disturbed, the sound of their wings resembles waves breaking on a troubled shore."

Natural Divisions.

There are some minor natural divisions in each of these main divisions. In the northern alluvial plain the western portion is an old formation, the general level of which is higher than the eastern portion. The latter, which is included in the Pharkiya *pargana* or Gogri thana, has an extensive block of high land in the north, and a low riverain strip to the south; but it is mainly a saucer-shaped depression, the centre of which is inundated during the rains by the overflow of the rivers and for the rest of the year is full of marshy hollows. In the rains this tract, which extends over some two hundred square miles, is a vast swamp so deeply flooded that not more than half the land is cultivated. In the dry season it is a wide prairie covered with a rank *pod* grass and the graceful pampas, together with an undergrowth of more succulent grasses, which afford abundant pasture for great herds of cattle.

The western portion is included in the Begusarai subdivision and is fringed on the north by a level upland tract, and on the south by the Gangetic riverain, where land is constantly being formed or washed away by the swift silt-laden current. To the north of the Begusarai subdivision is a large but shallow lake called the Kabar Tal, and east of the latter are large areas of grass jungle intersected by rivers and swamps. Round the lake, to the north of the river Gandak, is a belt of high land containing several indigo factories. To the south of the Gandak there is a strip, about three miles broad, which is liable to flood and is inundated every year. Further south, along the Hajipur-Katihar extension of the Bengal and North-Western Railway, the country is densely populated and well cultivated. Indigo used to be grown in the western portion of this tract, but owing to the competition of chemical dyes ordinary country crops have taken its place.

In the country south of the Ganges there is an alluvial strip of land stretching along the bank of the river and bearing heavy *rabi* crops. Immediately south of this alluvial belt the Kharagpur Hills form a distinct watershed, the country to the west being drained by the Kiul, and that to the east by the Man and other streams. To the north-west, in the angle between the South Bihar Railway and the East Indian Railway, from Lakhisarai to Barhiya, there is a wide level plain, in which a few detached hills rise abruptly from a level expanse of rice-fields; part of this tract consists of the flood area of the Halahar or Harhohar river, and is characterized by a heavy soil. To the south of the South Bihar Railway line is a wide, almost treeless plain, comprised in the Sikandra thana; and this is separated from the Kiul valley to the south by a block of hills known as the Gideswar Hills.

South of the Kharagpur Hills there is a stretch of undulating country extending to the borders of Chakai and Hazaribagh, which comprises *parganas* Parbatpara and Chakai, and a part of *pargana* Gidhaur made up of three large *taluks*, viz., Mahapur Kalan, Dumri and Mahesri. *Pargana* Chakai in the extreme south is an upland plateau encircled by hills and still largely covered with jungle. It is separated from the undulating tract to the north by a range of hills called the Batia Hills stretching in a long curve from Barwe, north of Simaltala, to the Hazaribagh and Gaya border.

HILL SYSTEM.

The hills of Monghyr comprise a number of low ranges and isolated peaks, outliers of the Vindhyan series, which enter the district from the south and gradually converge towards Monghyr town, where they dip under the Ganges. The most extensive range is known as the Kharagpur Hills, which form an irregular triangular block extending from near Jamalpur to the Jamui railway station. They consist of a number of steep ridges rising from the low ground on all sides, with scarped faces of massive quartzite in places; they are of irregular formation and do not run in any uniform direction. Generally speaking, the range is a bold and striking mass of rocks covered for the most part with jungle; but it contains valleys with patches of cultivation and several hot springs, of which the finest are those at Bhimbandh, though those at Sitakund and Rishikund are better known. Near the south-western fringe of these hills is Sringirikh, a peak said to have been the hermitage of the *rishi* Sringa and a noted place of pilgrimage. There are several peaks rising to a height of about 1,500 feet, and the highest point is Maruk (1,628 feet above sea-level), a table-topped hill, covered with forest and crowned with a deep layer of laterite. To the north of this range are low jungle-covered spurs approaching within a short distance of Monghyr.

To the south-west is another block of hills, which are known locally as the Gidheswar Hills from a peak of that name, but are referred to in geological works as the Gidhaur Hills. These hills are a continuation of the hills in the Nawada subdivision of the Gaya district, and cover an area of about eighty square miles, forming a compact cluster between Khaira and the western boundary of Monghyr. They rise sharply from the plain, but in most places there is a belt of jungle along their northern face before the actual ascent begins. To the east there is a fine cliff overlooking Khaira and the Kiul river, and the range falls away to the south into the rocky valley of the Kiul. On the south, in the village of Sakdari, there is a spring called Panchbhur, which is surrounded by precipitous walls of rock. The highest point of the range is at Ekgora (1,813 feet).

To the south a broken semi-circular range extends from near Bishunpur on the west to Simaltala on the east, separating the Chakai plateau from the rest of the Jamui subdivision. On the extreme west of this range is a high hill, named Satpahari, scarped on its northern face, beyond which the Kiul river breaks through the range by a narrow gorge. The hill is 1,806 feet above sea-level and 1,200 feet above the country at its base. There is also a small range of hills in the level alluvial plain near Sheikhpura, which are practically bare of vegetation. They rise somewhat abruptly on the south, while on the north, where they overlook Sheikhpura, the crags are almost precipitous. The range is intersected by several miniature passes, over which the roads are carried. There are also small isolated stony hills south of Sheikhpura, and some hills of fair size on both sides of the Lakhisarai-Jamui road south of Titar Hat in the plain to the west of the Kiul river.

Hot springs.

The Kharagpur Hills contain several hot springs, which are more fully described in Chapter XIV. Mr. V. H. Jackson classifies them into three distinct groups, in each of which the individual springs, though they may be many miles apart, so closely resemble one another that they are probably connected with, and derive their warmth from, the same under-ground water-system. The finest group, consisting of those at Bhimbandh, Malnipahar, and Karmandhari, with maximum temperatures of from 149° to 151°, are evidently associated with faults close to the southern border of the hills.

The second group, consisting of Rameshwar Kund near Kharagpur, Rashikund, and Bhurka, all rise at the foot of the eastern boundary and are much cooler, with maximum temperature of about 114°. The best known springs, Sitakund near Monghyr, and those in its immediate vicinity, are further north-east, and though they rise in the plain at some distance from the main mass of the hills, they are distinctly hot, with a maximum temperature of 139°.

As in all springs rising from silicious formation, the water is exceedingly soft and pure. According to Schulten's analysis of the Bhimbandh, Karmandhari, Rameswarkund and Rishikund springs, the total solids dissolved range from 5·4 to 7·5 parts in 100,000, more than half in each case consisting of merely silicious matter.

The gases which usually bubble up with the water have been examined at Bhimbandh and Malnipahar by Mr. Jackson, who considers them to be merely air dissolved by rain and modified by subsequent underground contact, because they consist chiefly of nitrogen with some carbon dioxide. Other inert gases may be present, but neither the gases nor the silicious deposit on stones in the springs are definitely radioactive.

Though the Chinese pilgrim, Hiuen Tsiang, who visited the neighbourhood of Monghyr in the first half of the seventh century A.D., records that "by the side of the capital and bordering on the Ganges river is the I-lan-no mountain, from which are belched forth masses of smoke and vapour, which obscure the light of the sun and moon", there are no apparent signs of such comparatively recent volcanic activity in the neighbourhood as this would indicate. Contrary to the general belief, the outflow of these springs is by no means constant. Their maximum volume, which they attain either during or shortly after the close of the rainy season, depends upon the nature of the monsoon; and Mr. Jackson considers that these variations in volume are sufficient to explain the changes of temperature which are noticed in all, though they are slight in the hottest springs.

Most of the hot springs are held in considerable repute by persons who live in their neighbourhood as potent remedies, especially for itch, ulcers, and other skin affections. An essential part of the process of cure consists in the preliminary worship of the presiding deity of the spring.

River System. The rivers of the district consist of the Ganges, its tributaries and sub-tributaries. To the north its principal affluents are the Burh Gandak, the Baghmati and the Tiljuga or Kamla, which enter the district from Darbhanga. They have comparatively narrow channels, and after a heavy fall in the sub-Himalayan tracts from which they debouch, frequently overflow their banks and lay a considerable area under water. To the south the principal rivers are the Kiul, the tributaries of the Kiul, and the Man; but there are also a number of hill streams, which come down in freshets during the rains, but subside as rapidly as they rise. For the greater part of the year they are sandy water-courses with little or no current, and are exhausted by the demand for irrigation before they reach the Ganges. The following is a brief account of the principal rivers.

Ganges. The Ganges has a course of about seventy miles within the district. It first touches it a few miles to the west of the Bachhwara railway station nearly opposite Barh in the Patna district, from which point it flows to the south-east in a long reach of thirty miles as far as Surajgarha, where it is joined by the Kiul. Leaving Surajgarha, it flows to the north-east describing a sharp bend on reaching the high land near Monghyr. It then turns almost due south for twelve miles, and next to the eastward for five miles till it reaches the boundary of Bhagalpur. In its course through the district the river is both wide and deep at all times of the year, and in the rains it spreads over the low-lying lands of *pargana* Pharkiya to the north for a distance of twenty miles from its ordinary bed. To the west, it is said to have had formerly a course ten miles to the north of its present channel. To the east, the river has several times shifted both to the east and to the west of the rock on which the Monghyr fort stands, alternately forming and washing away large areas of *diara* lands; but since the earliest times of which any record exists, it has washed the base of the rock immediately to the north of the fort. The largest areas of alluvial deposit formed by changes in the main channel are comprised in the Government estates of Kutlupur to the west and Binda *diara* to the east of Monghyr town. The *diaras* on the southern bank are very extensive, so much so that the breadth of the river some miles above Monghyr, when in flood, averages from six to eight miles.

Burh Gandak. The Burh (literally Old) Gandak, or Little Gandak, runs due south along the north-western boundary of the Begusarai subdivision for a distance of about eight miles from near Rusera in the Darbhanga district. It then turns to the east, entering the district at a village called Akaha, and follows a winding course through the Begusarai subdivision until it flows into the Ganges by the town of Khagaria, a few miles below Monghyr. It is navigable all the year round, for large boats during the rains and for small boats at other times.

Balan. The main tributary of the Burh Gandak is the Balan, which flows from the Darbhanga district and then takes a south-easterly course, joining the Burh Gandak at Umedpur near the Chiria Bariarpur police outpost. It has one small tributary, the Bainti, a stream debouching from Darbhanga, which falls into it near its junction with the Burh Gandak. The Balan has a considerable volume in the rains, and is navigable by small boats throughout the year.

Baghmati. The Baghmati enters the district a little east of Garhpura in the north-east corner of the Begusarai subdivision and then flows east into the northern portion of the Monghyr subdivision near Bahadurpur. It traverses that subdivision, pursuing a winding but generally easterly direction, till it flows into the Tiljuga near Chautham. It has one tributary, the Chandan, which is fed mainly by the Kabar Tal.

Tiljuga. The Tiljuga, also called the Kamla and, in its lower reaches, the Ghagri, enters the Gogri thana from Darbhanga a few miles north of Mohraghat. It flows south-east to Chautham, where it receives the waters of the Baghmati; and the united stream then continues to the boundary of Bhagalpur under the name of the Ghagri. Near Ramnagar it is joined by the Katni, which enters the district about ten miles to the north near Kasnagar and is itself a combination of three streams called the Telawe or Talaba, Parwan and Loran. The Tiljuga is navigable all the year round by boats of considerable size.

Kiul. The principal river to the south of the Ganges is the Kiul. It rises in the Kharagdiha thana of Hazaribagh, and after forming the boundary of that district for a short distance, enters Monghyr through a narrow gorge near the Satpahari hill. It runs at first eastward close to the southern face of the Gideswar Hills, but turns northward at their eastern extremity and passes one mile east of the town of Jamui. Two miles south of Jamui it is joined by the Barnar, which rises on the borders of Chakai and Hazaribagh, and two miles below this point it receives the Alai, a hill stream, which, like other hill streams in the south, dries up in the hot weather. Opposite the Jamui railway station it is joined by the Anjan, which drains the northern portion of the Jamui subdivision. It then flows north-east up to Lakhisarai, and is joined a few miles north of that place by the Halahar (or Harhohar), a continuation of the Sakri river. After this it turns due east and finally falls into the Ganges near Surajgarha. Until it meets the Halahar, the Kiul has a broad sandy bed, and in some places is as much as half a mile wide, though it contains very little water in the hot weather. It is spanned by a large railway bridge between Kiul and Lakhisarai.

The Anjan rises in the Anjan Hill to the north of the village of Barhat in the Mallepur *taluk*, and after passing through that *taluk* falls into the Kiul near Bariarpur. It has been dammed up in Barhat, and its water is diverted into an irrigation channel known as the Belia Nali, which irrigates several villages. The river receives several tributary streams, viz., the Jamkhar, the Bajan, and the Chhuria, which has a sub-tributary, the Kairwar. Popular tradition states that Anjani gave birth to Hanuman on the hill in which the river has its source. Anjan.

The Ajai, one of the larger tributaries of the Bhagirathi, has its source in the extreme south. It owes its origin to the confluence of a number of small streams, but before it leaves the district is a fairly large river. It eventually debouches into the Bhagirathi near Katwa in the Burdwan district. Ajai.

The Man river rises in the Kharagpur Hills, not far from the Bhimbandh springs, and pursues a winding north-easterly course to the east of those hills till it debouches in the Ganges near Ghorghat. The lower reaches, however, contain but little water, for a great dam has been built about two miles south of Kharagpur, where the Man runs through a narrow gorge. To the south the gorge widens out into a valley, hemmed in by low but abrupt hills, which the dam has converted into a large reservoir, from which water is drained off to irrigate the adjoining country. Man.

Lakes and marshes.

The country north of the Ganges abounds in marshes, several hundreds being enumerated in the Pharkiya *pargana* (Gogri thana) alone. Their formation is generally peculiar. The banks, which are covered with wild roses and the lantana, or wild sage bush, are high and abrupt, and they would seem to owe their origin to the diversions of the great rivers of the district. They are filled annually by the floods of the Ganges or its Himalayan affluents, and during the rains abound with snub-nosed crocodiles. They are also full of fish and form the chief source of the Monghyr fish supply. Rice is sown on the edges of the shallow marshes, and the deeper ones are frequented during the cold season by wild fowls, geese, ducks and waders in extraordinary numbers. An idea of the immense flocks found on these marshes may be gathered from a visit to one of their roosting places near Sakarpura in *pargana* Ballia, where the egrets, herons, ibises, coots, cormorants and snake-birds (*blotus*) assemble from all the country round to roost on the marsh-oak myrtle trees (*Barringtonia*), which stand out of the water. About sunset the horizon becomes streaked with flocks, many of them extending a mile in length, and by the time darkness has set in, hundreds of thousands of birds have taken up their quarters for the night.

These marshes also yield a vast quantity of shells, such as those of the fresh water mussel (*unio*) and the marsh snail (*ampullaria*), from which lime is prepared. In the cold weather it is a common sight to see two or three parties of old women in a state of semi-nudity gathering in the harvest which the retiring water has left them. The first party is picking up the shells of the marsh snail and the little glass snail. The snails have nearly all been picked out by the shell ibis, but there are enough left to afford a meal, and the empty shells, consisting of nearly pure lime, are sold to traders in the neighbouring bazar. A second party will be seen digging up the underground creeping stems of a water-lily or the sedge bulbs, called in the vernacular *chichor*, which are eaten to give a flavour to the snails and crabs, which the third party are engaged in securing.*

Further to the west there is another chain of marshes all along the north-east of the Begusarai subdivision, of which the most important is the Kabar Tal. This is a large shallow

* *The Natural Productions of the Monghyr District north of the Ganges*, Statistical Reporter, March 1876.

lake extending over an area of nearly seven square miles, a portion of which is always under water, but the remainder dries up in time for the sowing of rice broadcast in the month of May, the crop being reaped in November.

South of the Ganges, permanent marshes are of insignificant size; but considerable tracts of country are flooded in the rains. The reservoir at Kharagpur, however, forms a beautiful lake, which has been described by a former Collector as " rivalling the renowned lakes of Killarney "

Forests. There is an area of several hundred square miles under forest towards the southern confines of the district and in the Kharagpur estate.* The principal tree is *sal* or *sakwa* (*shorea robusta*), but, owing to the absence of any system of forest conservancy, the giants of the forest have long since disappeared. Even when the East Indian Railway was being constructed over half a century ago, scarcely any large trees could be found capable of supplying sleepers. Another conspicuous tree in these hills is the *kend* (*dispyros melanoxylon*), the black heart wood of which forms the ebony of commerce and is in great demand among the Monghyr carpenters. Another valuable timber tree is the *aunla* (*phyllanthus emblica*), the fruit of which is made into a chutney and also into a hair-wash.

The gorgeous flowering *palas* (*butea frondosa*) abounds and is even more useful. Its wood is used for fuel and its coarse fibrous roots for caulking boats. The lac insect breeds on its small branches and petioles, and it affords a valuable gum and yellow dye of considerable permanence. When in blossom in March, the tree is a blaze of scarlet flowers, which have earned for it the picturesque name of the "flame of the forest". The allied species called *chihunt* (*butea superba*) is also abundant, and its flowers are equally splendid. *Kachnar* trees (*bauhinia purpurea* and *variegata*) are numerous, and during the cold season lighten up the woods with their beautiful flowers. The gigantic creeping *chehar* (*bauhinia vahlii*) does great damage, for it climbs over the tallest trees and kills them in its deadly embrace; but it is economically useful, for ropes are made from its bark. Other common leguminous trees and bushes are the tamarind, the *amaltas* or Indian laburnum (*cassia fistula*), and many thorny

* This account of the botany of Monghyr has been compiled mainly from *The Natural Productions of the Kharagpur Hills in Monghyr*, an article published in the Statistical Reporter, 1876.

mimosas. Among other important members of this family several species of *terminalia* may be mentioned, including the *harra* (*terminalia chebula*), the fruit of which yields the myrobalans of commerce, while the wood makes excellent fuel.

Malvaceæ and *sterculiaceæ* are also numerous. The former species includes the *simal* or red cotton tree (*bombax malabaricum*), various kinds of *hibiscus*, all yielding strong fibres, and the *ban kapas* or wild cotton. The latter species is represented by *mogul* or *karaunji* (*sterculia urens*), the fibrous bark of which is almost unbreakable. In December it is conspicuous for its white trunk and leafless branches, which have earned for it the description of a spectre-like tree. Special mention must also be made of the forest *sabe* (*Ischoemum angustifolium*), a species of grass found on the hill tops. Hundreds of persons gain a livelihood by collecting and twisting it into string, and at least three-fourths of the string used in the district is made from it. Resin-bearing trees are common, especially the Indian frankincense or *saleh* (*boswellia serrata*), which grows to a large size, and is one of the few trees allowed to reach maturity. In the cold weather it is covered with pink flowers which attract hundreds of the yellow-tip epicharis, almost the only butterfly one meets in these arid jungles. There are also several species of fig trees, conspicuous among which is the *gular* (*ficus glomerata*), the fruit of which the forest tribes share with the birds and bears.

Mutilated as they are, these forests are still of great value to the people of South Monghyr. The trees which they fell for timber or fuel supply them with the means of getting their daily bread. Their houses are built with the bamboos which grow on all sides, and are thatched with coarse grass, which is to be had for the trouble of cutting it The *kachnar* creeper, the *mogul* bark and the *sabe* grass supply them with ropes and string. In seasons of scarcity the tubers of the wild yam family, supplemented with a little rice, furnish their daily fare. When they are sick, they send to the forests for medicine; and the plant sought for will depend on whether the part afflicted is the head or stomach, hand or eye.

Useful as most of the trees undoubtedly are, there is no tree which can be compared to the *mahua* (*bassia latifolia*), which yields food, wine, oil and timber. From its flowers the common country spirit is distilled, and, whether fresh or dried, they furnish the poorer classes with wholesome food; from the fruit is pressed an oil largely used for the adulteration of *ghi;* and the tough timber is used for the naves of cart-wheels. "Unlike mango trees, so uncertain in their yield of fruit, however plentiful the blossom may have been, the chief, though not the only, value of the *mahua* depends on the succulent petals of the flower, which cover the trees from year to year, apparently regardless of favourable or unfavourable seasons, so pregnant with weal or woe to other plants. The flowering season, in March, is a great season for feasting among the humbler members of creation. Birds, squirrels and tupaias feast among the branches by day; whilst men women and children sweep up and carry home the petals, which fall around in great profusion. Nor does the feasting end with the day: bears, pigs and deer have their turn during the night, and many of them fall a victim to their fondness for the *mahua* flower, being shot by the bullets of the foresters concealed among the branches overhead."*

North of the Ganges the older rocks are concealed by the alluvium of the Gangetic plain, but south of the river the level of the land rises rapidly and the older rocks soon appear, first as more or less disconnected hill groups, and further south as a continuous uninterrupted outcrop. These rocks consist of the oldest system recognized by geologists, that known as Archæan. They include:—(1) a vast series of crystalline rocks of varied composition, including granitic and dioritic gneisses, hornblende and mica schists, epidiorites, crystalline limestones and many other rocks collectively known in the geological nomenclature of India as Bengal gneiss; (2) another very ancient series consisting of highly altered sedimentary and volcanic rocks, including quartzites, quartz schists, hornblendic, micaceous, talcose and ferruginous schists, potstones, phyllites, slates, etc., which form an assemblage very similar to that which has received the name of Dharwar schists in Southern India; and (3) vast granitic masses and innumerable veins of coarse granitic pegmatite, intruded amongst both the schists and the Bengal gneiss. GEOLOGY.

The Bengal gneiss occupies principally the southernmost part of the district. The ancient stratified series assimilated with the Dharwars forms several hill groups situated between

* *The Mahua Tree in Monghyr,* Statistical Reporter, December 1875.

the southern gneissose area and the valley of the Ganges: these are the Kharagpur Hills, the largest of the hill masses situated south of Monghyr and east of Lakhisarai, the Sheikhpura Hills and the Gidhaur range, situated respectively west and south of Lakhisarai. The rocks of the Gidhaur range are highly metamorphosed by innumerable veins of coarse granitic pegmatites, which are of great economic importance on account of the mica which they contain, and constitute the eastern portion of the great mica belt of Bengal. The coarsest grained, and consequently the most valuable, pegmatites are the comparatively narrow sheets which intersect the schists of the metamorphosed stratified series. The larger and more uniform, comparatively fine-grained intrusions are valueless so far as mica is concerned, though they belong to the same system of intrusions. On account of its habit of weathering in the shape of large rounded hummocks, the rock forming these more massive intrusions has been often described under the name of dome-gneiss, which, more accurately, should be dome-granite. The rocks of the Kharagpur Hills are not nearly so much altered as those of the Gidhaur range. The strata originally constituted by shales, which, in the latter range, have been transformed into schists, are only altered to slates in the Kharagpur Hills. These slates, which are regularly cleaved and of fairly good quality, are quarried to a certain extent.*

Minerals are confined to the tract lying south of the Ganges. Galena, a sulphuret of lead, containing a small quantity of silver, is found in the hill tracts of *pargana* Chakai, and minium or protoxide of lead in the beds of the Kharagpur hill streams. Mica occurs in the belt of schists and gneissose granite which stretches north-eastwards from the Gaya district to near Nawadih (Jhajha) on the East Indian Railway. Iron ores are found in the schists of the Kharagpur Hills, and in several places cchreous ores are employed as pigments. Slates are quarried at Basauni near Dharahra, and stone quarries are also worked. Felspar fit for the manufacture of porcelain occurs in abundance in the south of the district. Corundum is obtained from the hills near Jamui, but the precious forms are not met with. Travertine is fcund near Gidhaur and in the Kharagpur Hills.

*** This account of the geology of the district was contributed by Mr. E. Vredenburg, Superintendent of Geological Survey of India.**

BOTANY.

There are few districts in Bengal so favourably situated for the study of botany as Monghyr. The alluvial and thickly populated plains north of the Ganges contain most of the trees and plants useful to man that are suited to the soil and climate, while in the hilly country to the south there are numerous forest trees and shrubs. Here ebony, *sal*, red cotton and other trees grow side by side, protecting from the sun's rays the ferns and humble creepers which grow below. Anyone who takes his stand on the table-topped Maruk, or indeed on any point in the hills, may count a hundred different species growing round him. But the most casual observer will at once detect the ravages made by the agency of man. Not only is the woodman's axe busy, and the underwood periodically consumed as firewood, but cows, sheep and goats are let loose in the woods, and pasturage quickly destroys vegetation. Indeed, in the Sikandra thana for some hundred square miles not a grove can be seen, and there are few trees except those planted along the roads or a line of palm trees along some tank or the boundary of a field. Notwithstanding, however, the denudation which goes on, Monghyr is still a well-wooded district, for round the villages many trees are allowed to survive for the sake of their fruit, while a botanical excursion in the hills will disclose a great variety of trees and plants, although few of the trees are of any large size.

The *sal* trees especially have suffered from indiscriminate felling. Directly they attain a size sufficient to form a *gol* or prop for a house, they are chopped down, carried off, and sold by their owners. The consequence is that other kinds of trees, whose wood is in less demand for fuel or for timber have obtained the upper land. The fibre-yielding *malvaceæ* are perhaps now the most conspicuous; their wood making indifferent fuel, it is hardly worth while to strip them of their bark, as the fibre of the cultivated members of the family is sold cheap in every bazar. The red cotton tree, the *karaunji* (*sterculia urens*), and the *ainthia dhamin* (*helicteres isora*) with its twisted pods are common; while the wild cotton and many other species of *hibiscus* appear as annuals or undershrubs. The leguminosæ are also common, including the tamarind, the *amaltas* (*cassia fistula*), the *palas* (*butea frondosa*), the *kachnar* (*bauhinia variegata*) and many acacias and mimosas. The *karjani* (*abrus precatorius*) is another noticeable member of this family, its vermilion seeds being conspicuous in the jungle during the cold season. The *saleh* (*Boswellia serrata*), belonging to the *meliaceæ* family, is found throughout the hills, and is less molested than almost any other tree, because its wood is of no value for timber, and as fuel it gives out such dense clouds of smoke that no one cares to burn it. It is a great ornament, however, to the forests, and in the distance resembles the English mountain-ash.

Next in numerical order come the *apocynaceæ*, such as the *dudh koraiya* (*Wrightia tomentosa*) and *dudhi* (*holarrhena dysenterica*), which appear everywhere in the hills, and are always conspicuous in the cold weather with their long follicles or pods winged with a tuft of silken hairs. These, when ripe, burst open, and are borne by the wind to any ground which may be unoccupied. The *rubiaceæ* are also well represented in these woods. The Nepal lilac (*Hamiltonia suaveolens*) is one of the few common plants with fragrant flowers which adorn the hills during the cold season. When this tree and the *Holmskioldia sanguinea* grow side by side, as they often do, they stand out, the one with blue or white scented flowers and the other with red blossoms, in beautiful relief against the background of dark green trees. The same family includes the *bhorkhorn* (*hymenodictyon excelsum*), which may always be recognized during the cold season by its brown capsules, containing winged seeds, hanging in clusters on the leafless branches. An account of the trees which are found in the Monghyr Hills would be incomplete without mention of the *terminalia*, which, if only given fair play, would grow into large trees and supply timber little inferior to *sal*. The *asan* (*terminalia tomentosa*) is common, but the *harra* or black myrobalan trees (*terminalia chebula*) are becoming very scarce in consequence of the activity of the woodman's axe. There is also the *kahua* (*terminalia arjuna*); and an allied species, the *dhao* (*anogneissus latifolia*), is very common, commanding as fuel by far the best price in the Monghyr market.

In the forest clearings there are a number of plants imported from America, such as Indian-corn, potato and tobacco. Here too are found the cactus, the Mexican poppy (*argemone Mexicana*), and the *sher-nui* (*Martynia diandra*), a weed which in a few years spread far and wide. The villagers assert that the tiger will not come into a field where it grows for fear of its prong-like seeds getting entangled in his coat. Among other noticeable plants which are cultivated in the clearings among the hills is a pigmy sunflower (*verbecina satina*). This plant marks the Santals' village, and is not found elsewhere in the district. It yields a delicate oil, which, with sesamum or *til*, also characteristic of forest clearings, is used to lubricate the coarse food which the woodmen eat. On the banks of the hill streams, particularly near the waterfalls among the Kharagpur Hills, in the course of the river Man, many beautiful flowers will be found during the cold season. Among these may be mentioned the *Holmskioldia sanguinea* with a blaze of red flowers setting off the blue *barlerias* which grow below, and the *porana paniculata* (called by Europeans the silver creeper or the bridal creeper, but by Indians the *burhi* or old woman), which covers the trees and rocks where it grows.

Of the trees found north of the Ganges, the great majority are mango trees, which are grown in topes or orchards in every village not subject to prolonged inundation. In good years the mango forms no inconsiderable portion of the poor man's food; and in times of dearth it is an important item in the fcod supply of the district. But the crop is uncertain, and the trees are troubled with parasites, which do them much injury, though the scarlet flowers add considerably to the beauty of the landscape. Besides supplying fruit and fuel, the mango groves are useful as pasturage grounds, giving shade to cattle in the hot weather and shelter during the winter. The next most abundant tree is the oak myrtle (*Barringtonia acutangula*), which abounds in the marshes of *pargana* Pharkiya and which supplies much fuel to the Monghyr market. It is known here as the *ijar*, and, although belonging to the natural order of myrtles, has the appearance of a stunted oak. It grows well in several feet of water, and is consequently a favourite roosting place for birds. The branches, which during the rains droop into the water, are also the resting place of fresh-water sponges known to the villagers as *phen*, or foam. The red cotton tree is one of the most conspicuous trees, particularly in the cold weather, when it is covered with large crimson flowers.

Some members of the great fig family are found in every village. The *pipal* (*ficus religiosa*) is most common, and then the banyan (*ficus bengalensis*), though in this district the large trees with several trunks are comparatively rare. The *gular* (*ficus glomerata*), with its fruit packed full of flies, and the *pakar* (*ficus infectoria*) are also frequent. The well-known *babul* (*acacia arabica*), with its sweet-scented ally the *guhiya babul* (*acacia farnesiana*), is common in hedges and waste places, and is a favourite nesting place of the turtle dove. Then there is the *sahora* (*streblus asper*), the twigs of which provide the Hindus with tooth-brushes; but it is looked on with abhorrence by Musalmans, because, according to tradition, it was the only wood which would take fire when the unbelievers wished to burn their saint Ibrahim Khalil Ullah. The *kachnar*, *nim* (*melia azadirachta*), *siris* (*albizzia lebbek*) and tamarind are fairly common; and the *jiyal* (*odina wodier*), with its cat-o'nine-tail panicles of flowers, attracts attention in every well-wooded village. The *sisu* (*dalbergia sissoo*) is planted freely, usually in rows on the borders of the fields, where their crooked trunks lopped of all the lower branches present a fantastic appearance. Willows (*salix tetrasperma*) may be seen on moist banks overhanging watercourses and marshes. They are known locally by the name of *bes*, and baskets are occasionally made from their twigs in Monghyr as in England. Palm-trees, including both the palmyra (*phœnix flabellifer*) and the date-palm (*phœnix sylvestris*), are found in plantations round the villages; but they are not so common as they are in the southern part of the district, where during the hot months thousands of gallons of toddy (*tari*) are consumed by a thirsty public and yield revenue to the State. Among other fruit trees may be noticed the *jaman* (*eugenia jambolana*), the jack or *kathal* (*artocarpus integrifolia*), and last but not least the *bel* (*angle marmelos*). To enumerate all the trees would require much space; those which have been mentioned are merely the most common and conspicuous.

The hedges which exist in the vicinity of the villages contain a large variety of plants, among which may be mentioned the *bagnai* (*capparis horrida*), the *madar* (*calotropis gigantea*), the *samalu* (*vitex negundo*), the *jasmine* (*chameli*), the wild castor-oil plant and many others. In the uncultivated parts of *pargana* Pharkiya by far the most conspicuous plant is the *koa* or wild rose of Bengal (*rosa involucrata*) covered with flowers, which for sweet scent is surpassed by few flowers in the world. The fields, when under *rabi* crops, have a wide variety of weeds, prominent among which is the Mexican poppy (*argemone Mexicana*), collected by the very poor for the sake of the oil extracted from the seeds.

Not the least valuable product of the Pharkiya *pargana* is thatching grass. It is grown on low land subject to inundation, which retains water too long to enable the raiyat to sow a cold-weather crop upon it. So extensive are these grass fields that they may be called prairies, and they could, if necessary, supply all the houses in Monghyr with thatch. The supply thus exceeds the demand; but large herds of cattle, chiefly buffaloes, which during the hot season are brought from all parts of the district, pick up a scanty living from the young shoots and undergrowth of *dub* grass and weeds. Very few flowers are found on the prairies, but the asparagus creeper is occasionally met with, and the lantana or wild sage is not uncommon. A creeping fig, with some of its leaves vinelike, frequents moist banks, and a parasitic orchid, which though not at first sight attractive, forms a beautiful study under the microscope.*

Writing in 1878, Mr. Lockwood remarked in his *Natural History, Sport and Travels* :—" Twenty years ago any one standing on the summit of the hills adjoining Kharagpur would have been surrounded by a dense and almost impenetrable forest where at sunrise peacocks and jungle-fowl were heard calling to their mates and where in the evening tigers, leopards and bears came to drink water of the little river Man, which, rising at the hot springs in the recesses of the hills, flowed, forming many beautiful cascades by the way, between the hills until it reached the plains near Kharagpur. The past twenty years, however, have effected greater changes in these hills than did the twenty centuries which had gone before. Directly the railway whistle was heard on the adjoining loop line of the East Indian Railway, the wild animals retired to happier hunting grounds; the peacocks and jungle-fowl followed, and soon the forest resounding with the woodman's axe gave way to fields of wheat, cotton, sesamum, and other fertile crops, such as may only be seen in India on virgin soil." ZOOLOGY.

In the years which have since elapsed, the number of wild animals and birds has decreased still further. One reason for this may be found in the gradual reclamation of jungle, which has forced them to migrate to uncleared forest in the south. Another may be found in the indiscriminate

---

* This account of the botany of Monghyr has been compiled mainly from an article, *The Forest Flora of Monghyr* published in the Statistical Reporter, March 1877.

slaughter of game, whether the larger wild animals or game birds, such as spur-fowl and jungle-fowl, by professional *shikaris* and the forest tribes in the south of the district. North of the Ganges too birds of gay plumage are killed by fowlers for the sake of their feathers. During the rainy season, when the egrets and purple herons put on their nuptial plumage, they are killed in considerable numbers in *pargana* Pharkiya for the feather market, while the scapular feathers of the plotus or snake bird mark it out for the same fate. At other seasons of the year kingfishers, bee-eaters and the Indian roller or blue jay are snared to meet the demand for their feathers. In spite of this destruction, Monghyr, with a large area of jungle-clad hills and watery swamps, can still exhibit a richer variety of animal and bird life than many other districts in Bihar.

Mammals. Tigers are still found in the Kharagpur and Gidheswar Hills, though scarce. They were responsible for 176 deaths in the three years 1900—2, these deaths being probably due to a family of man-eaters ranging between this and the Gaya district, where also they killed a number of human beings. Again, in 1908 one man-eater was reported to have killed 16 persons in two months in the neighbourhood of Jamui. The extermination of these brutes is no easy task, as the jungles which they frequent are very extensive, and so thick with under-scrub, that the sportsman cannot proceed on foot with any chance of success, while the passage of elephants is impossible. Leopards are much in evidence, and numerous cases are reported of their killing dogs, goats and cattle even within 15 miles of Monghyr. Black or sloth bears (*ursus melursus*) are found on all the larger hills and in the jungle surrounding them, where they can feed on white-ants, plum and other fruits, and drink from the hill streams. Even here, however, they are not so plentiful as even ten years ago, because they are easily shot down by the Santals, when they come to feed on the fallen flowers of the *mahua* tree. They will dispute the possession of these with the peasants who try to gather them—often with fatal results to the latter. Hyænas are found in the hills, and wild cats are common. The civet family is represented by the large Indian civet (*viverricula malaccensis*). The dog family include wolves, the familiar jackal and Indian fox. The wild dog is met with in the hills, but is becoming rare. There is one species of badger not uncommon in the same tract—the Indian ratel (*mellivora indica*). Among insectivorous mammals the tupaias or tree shrews call for special notice. They are arboreal animals looking like a cross between a squirrel and a rat, but are easily distinguished from the former by their ears and teeth. Rare animals elsewhere, they are still found in the Kharagpur Hills.

There are several species of *chiroptera*, including frugivorous bats, the long-armed bat and long-tailed bat, and other insectivorous bats, known to the villagers by the generic name of *chamgudri*. Rodents include porcupines, hares and the palm squirrel (*sciurus palmarum*), a misleading title as it is far more frequently seen in fruit gardens than on palm trees. Two species of monkey are found in the district, the long-tailed *langur* (*semnopithecus entellus*) and the shorter tailed Bengal monkey or *bandar* (*macacus rhesus*). The former is found to the south of the Ganges, and the latter affect certain localities from which apparently they never migrate. At Monkey Island, on the Kabar lake, they are particularly numerous, being venerated by the priests of the Hindu temple. In the year 1793 a yearly grant was made by Government for the express purpose of feeding the monkeys and lighting the temple on the island; but in 1852 the grant was discontinued, owing to its being misappropriated by the priests. The Ungulata are not numerous in Monghyr. There are, however, several species of deer, including the noble *sambar* (*cervus unicolor*), chital or spotted dear, ravine deer and barking deer, which are found in the south, though not in any great number. *Nilgai* (*boselaphus tragocamelus*) are also occasionally met with. Wild pig are numerous in the jungle at the foot of the Gidheswar Hills and are often trapped in pits by *shikaris*. Last among mammals may be mentioned porpoises, which are found in the Ganges.

Birds.

Game birds other than water-fowl are comparatively rare in this district owing to a variety of causes. In the south, where there is ample cover in the jungle-clad hills, the Santals and others are constantly shooting them down for the pot. Cats, foxes and other vermin are even more destructive, and there is a certain scarcity of suitable food, for edible berries and nectar-bearing flowers are rare in the forests. The birds consequently prefer the cultivated fields and gardens; and the silence of the woods is proverbial among those who visit them. Still there is a fair variety of species, though their actual numbers are small. Pea-fowl are found in the southern hill jungles, particularly in *pargana* Chakai, and are still often to be seen in different parts of the Kharagpur Hills. Being sacred to Kartik, the younger son of Mahadeo, they are rarely molested by Hindus. The jungle-fowl (*gallus ferrugineus*) is met with in the same jungles, and the red spur-fowl is occasionally seen. Grey partridges are abundant in scrub jungle south of the Ganges; the black partridge or francolin affects the grass prairies in the north of the district; and the *kyah* or marsh partridge is found in the low lands of *pargana* Pharkiya. The black-breasted quail breeds in this district, but is not nearly so abundant as the common quail, which migrates in the spring to breed beyond the Himalayas. It consequently has a better chance in the struggle for existence than it would have here, where so many enemies in the shape of cats, foxes, hawks and snakes, are ready to devour it and its eggs. Bustards are rarely seen.

The courier plover is found chiefly on the alluvial *diara* lands bordering the Ganges, and flocks of swallow plover may be seen hawking for insects on the banks of rivers. Other species of plover are common—the golden plover during the cold weather and the spur-winged plover at all seasons; while the Norfolk plover is frequently seen among the rocks. Cranes are cold weather visitors. Snipe are not so numerous in Monghyr as in Lower Bengal. The common and pintail species appear to be nearly equally distributed, but painted snipe and jack snipe are rare. God-wits appear on the marshes in *pargana* Pharkiya during the cold season, as well as curlews and whimbrels; and the graceful avocet may be occasionally seen along the banks of the Ganges. Among the stints and sand-pipers met with in the district may be mentioned the ruff, the little stint, the spotted, green, and common sand pipers, which appear in large numbers during the cold season, while the green-shank and red-shank are found on every marsh. The black-winged red-legged stilt (*himantopus candidus*) is another winter visitor. Nearly all the wading family are migrants, retiring beyond the Himalayas to breed.

Both the Indian jacanas—the bronze-winged and pheasant-tailed—remain all the year round in the marshes to the north of the Ganges. They are marsh birds with long toes and claws, which enable them to run over the floating leaves of water lilies and other plants. The coots are also well represented in the marshes, conspicuous among them being the purple species. Several kinds of rail also frequent the same localities, but not in such numbers as coots and water hen. Storks are found throughout the district. The adjutant is rarely seen; but the white-necked stork is very common during the winter months. Herons are plentiful in the reedy swamps towards the north, and during the rainy season the small pond herons, or paddy birds, as they are familiarly called. Bitterns do not appear to be so common here as in other parts of Bengal; but the night heron (*nycticoarx griseus*) abounds, and its familiar call may be heard on still evenings as it passes to its feeding-grounds. The spoonbill is occasionally seen in the Pharkiya marshes, and there are several species of ibis.

The marshes in the north of the district form the home of myriads of geese and ducks during the cold season. The following species have been observed:—the grey goose, barred-headed goose, white-headed goose-rail, the whistling teals, sheldrake, ruddy sheldrake, shoveller, pink-headed duck, gadwall, pintail duck, widgeon, teal, gargany, red-crested pochard, red-headed pochard, ferruginous duck and tufted duck. The crested grebe is found on most of the marshes, where its diving powers excite the wonder of all who witness them, for it can travel under water almost as fast as it can fly in the air, a few seconds sufficing for it to dive and reappear at a distance of several hundred feet. Terns also abound, especially the whiskered tern and gullbilled tern. The curious Indian skimmer, with its razor-like lower mandible much longer than the upper, may be seen skimming over the Ganges everywhere. Pelicans, or river sheep, as the villagers call them, are also seen occasionally swimming in the rivers and marshes, or soaring far overhead. Cormorants are common in congenial localities; and the nearly allied Indian snake-bird (*plotus melanogaster*) pays dearly for the beauty of its scapular feathers, for during the breeding season the *shikaris* pursue it as eagerly as they do the egrets.

Birds of prey are fairly common, including three kinds of vulture, viz., the common vulture, the black vulture and the Egyptian or scavenger vulture. The Indians look on them with abhorrence, and some of them will even pull a house down if a vulture alights on it. Of the falcons there may be mentioned the peregrine, a cold-weather visitor, and the *saker* (*falco cherrug*). The common sparrow-hawk is also found. Monghyr is rich in fishing eagles. The osprey frequents the Ganges and the large marshes north of that river. The white-tailed eagle is equally common, breeding in the district. Harriers are plentiful during the cold season, hawking over every field in search of small birds and lizards, ortolan and quail being especially marked out as their quarry. Owls are not very numerous, but the rock owl is found.

Swallows, martins and swifts are numerous. The European swallow abounds during the cold season, while the wire-tailed common swallow may be seen occasionally in the Kharagpur Hills. Night-jars are also found throughout the hilly tracts. The common bee-eaters are numerous at certain seasons, and the beautiful Indian roller, commonly called the "blue jay", is one of the most familiar birds in this part of the country. Kingfishers are not so common here as in Bengal, but during the cold weather they may be seen hovering over tanks in search of fish. Hornbills, also wrongly called toucans, are not uncommon; the Malabar pied hornbill has been met with in the Kharagpur Hills. Paroquets are common and do a certain amount of damage to the crops. There is no great variety of woodpeckers, by far the most common species being the golden-backed woodpecker (*brachyternus aurantius*). Barbets are met with everywhere, especially the little copper-smith, so called from its monosyllabic metallic call resembling the noise made by smiths in hammering copper vessels.

Of the cuckoo family there are not many species. The European cuckoo comes during the cold weather, but returns to the hills during the breeding season, and its familiar call is rarely heard. The pied cuckoo is heard everywhere during the rains, and the hawk cuckoo and *koel* abound at all seasons. One species of the hawk cuckoo is known as the brain-fever bird from the monotonous repetition of its call note in the hot weather. Coucals or crow-pheasants are plentiful, and *sirkir* (*taccocua*) are met with in the Kharagpur Hills. Honey-suckers, or humming-birds, as they are generally called by Europeans, are common; and so are nut-hatches during the cold weather. The hoopoo is found in every garden, and is not molested by Muhammadans, as it is supposed to have been a great favourite with Solomon. Three species of cuckoo shrike are found and one or two species of shrike, the species known as *lanius cristatus* being the first migrant to appear at the commencement of the cold season. The beautiful minivets are cold-weather visitors, and may be seen in flocks searching for insects in almost every grove. The Drongo shrike or king crow is also seen everywhere, while the Paradise flycatcher is not uncommon in well-wooded country.

Thrushes are poorly represented, but the orange-headed ground thrush and the blue rock thrush may occasionally be seen during the cold weather. Babbling thrushes, on the other hand, are fairly represented; about seven are usually seen together, hence the name *satbhai,* or seven brothers. Bulbuls are not so numerous as in Bengal; they are caught by the *shikaris* for the sake of their plumage and their value as fighting birds. Orioles abound in every grove, where their beautiful plumage is a striking feature during the spring. The Indian redstart and blue-throat are common during the cold weather, when the ruby-throat is also occasionally seen in the *rabi* fields. Indian tailor-birds are plentiful and so are reed-warblers, which are particularly numerous on the Kabar lake. Tree-warblers are frequently seen in the cold weather, as well as wagtails, which are all migrants, leaving before their pairing season in the spring, and returning at the first approach of cold weather. Their arrival is always welcome, as it denotes that the thermometer will fall below 80° at sunrise.

The field wagtails are amongst the most numerous of the cold-weather birds, and the same may be said of their cousins the pipits. The grey titmouse (*parus cinereus*) also occasionally appears during the cold season. There are several species of the family *corvidæ,* including the corby and the ubiquitous common crow. The Indian magpie is another of the most familiar birds. The starlings are well represented by the *mainas,* several species of which remain all the year round. The European starling is common during the cold season, coming in large flocks, and being nearly always joined by the *mainas.* The bank *mainas* are also abundant along the Ganges, breeding on its banks. The *fringillidæ* or finch family are not so well represented as one would expect, considering the abundance of food suitable to them, but the weaver-bird is common, and its curious nests may be seen in many villages. The *munias,* known locally as *lals,* are favourite cage-birds, thousands being captured every year. The Indian house-sparrow is ubiquitous. The rosefinch, or *tuti*, is seen during the cold weather. The curious little finch-lark (*pyrrhulauda grisea*) is common in the fields, as well as the skylark, which is particularly plentiful in *pargana* Pharkiya. The green pigeon is frequently met with, as also the blue rock pigeon, which occasionally enters verandahs to build its nest on the cornices. Turtle-doves are found in almost every grove; and the beautiful ground-dove, with emerald-like wings, is sometimes seen.

Fish.

Fish are abundant in Monghyr, being found in large numbers in the Ganges and its affluents, and in the marshes to the north, which are replenished yearly from the river floods. The principal species are members of the great carp family, including the well-known *rohu* (*labeo rohita*), *kalbaus* (*labeo calbasu*), *mirgal* (*cirrhina mirgal*), *katla*, (*catla buchanani*) and many others. They grow with extraordinary rapidity notwithstanding their number, as an instance of which it may be mentioned that some years ago when a tank at Monghyr was dragged, 800 carp averaging two pounds each were landed, though they were only fry put in during the preceding rains. The siluroids, or scaleless fish, are also represented by the huge *gunch* (*bagarius yarellii*), *the silan* (*silundia gangetica*), the *tengra* (*macrones tengara*), the *boali* (*wallago attu*), the *bachua* (*eutropiichthys vacha*) and the *pangas* (*pangasia buchanani*), all of which afford excellent sport to the angler. Then there is the curious hunch-back or *moh* (*notopterus kapirat*), which attains enormous dimensions, and others too numerous to mention. Among small fish found in the Pharkia jhils may be mentioned the curious *koi* or climbing perch (*annabas scandens*), the beautiful prickle-sides (*trichogaster fasciatus*), and several kinds of tetrodon, or balloon fish, which has the power of inflating itself like a balloon and erecting its spines as a means of defence. *Hilsa* (*clupea ilisha*) pass up the Ganges in the rains, that being the season when they go up the rivers to spawn; and the river Man contains *mahseer* (*barbus tor*), locally known as the *kajur*.

The pools below the waterfalls along the latter river are tenanted by a little fish, which the wocdmen declare to be the young *tengra*. When the flood comes this little fish find it very difficult to hold its own against the stream; but nature has provided it with a sucker, which enables it to fasten itself to the rocks and wait securely until the flood has passed. Another fish found in the pools is the little emerald-streaked *barilius rerio*, which remains a pigmy all its life, hiding itself under the rocks when the floods come down. The woodmen delight in fishing in the hill streams. They use a fresh-water shrimp for bait, and, besides fish, they catch and eat the little Indian crab, which is plentiful wherever there is water.

Crocodiles abound in the north of the district. The common fresh water crocodiles (*crocodilus palustris*), which are known as *magar* or *bochboch*, are very numerous in the Tiljuga river, and during the rains in the neighbouring inundated marshes. A hundred of these brutes may be counted during the cold season basking in the sun along a single reach of the river. The long-snouted fish-eating *gharial* (*gavialis gangeticus*) chiefly affects the Ganges, and at sunrise several may be seen on every sandbank. Small specimens are not infrequently caught in the fishermen's nets, and are eaten by the lowest classes. Fresh-water turtles abound. **Reptiles.**

The most remarkable lizards in the district are the *varanidae* or *gohsamps*, the iguanas of Europeans; three species are found,—*hydrosaurs salvator*, the monitor (*varana dracaena*), and *empagusia flavescens*. The monitor, which may be distinguished by the position of the nostrils half-way between the lip and the eye, is supposed, without reason, to be very venomous. There are also many species of small tree-lizards, among which is the dreaded though perfectly harmless "blood-sucker"; and the chameleon is not uncommon. Among other thick-tongued lizards may be mentioned the wall *gecko*, which is a familiar sight in houses.

Snakes are numerous in Monghyr district. The cobra is by far the most common snake, and next to it the *karait* (*bungarus cæruleus*). The rock snake (*python molurus*), the boa-constrictor of Europeans, is found on the hills. The *raj-samp* (*bungarus fasciatus*) and Russell's viper are found, though rarely. Vipers are occasionally found. The *sankra* (*lycodon aulularia*), a pretty little harmless snake believed locally to be very deadly, frequents, as its scientific name implies, dwelling-houses and other buildings. Another snake, rare in museums, is not uncommon here, viz., *ferrania sibboldii.* The little *typhlops*, which resembles the earthworm, is common. It is known as the *thilia samp*, the meaning of which is said to be that the poisonous effects of its bite can only be removed by the speedy application of an earthen vessel (*thilia*) full of oil.

CLIMATE. The climate of Monghyr may be said to form a medium between the dry parching heat of Hindustan and the close moist atmosphere of Bengal. The seasons are the same as in other parts of Bihar. The hot weather commences in the beginning of March and continues till the middle of June. At this period of the year the westerly winds blowing across the arid sun-baked plains of Central India cause high temperature combined with low humidity. The heat is often intense, and the use of *tattis* is necessary to reduce the temperature of the houses; but the dry heat of the day is somewhat modified by a fresh wind blowing through the night. In some years, however, the hot weather is more like that of Eastern Bengal, there being a damp and very oppressive heat, with treacherous east winds and hardly any west wind. There are occasionally violent gales and storms during this season of the year, which sometimes bring up rain, and if rainless give rise to whirlwinds and blinding clouds of dust. In April and May thunderstorms, accompanied by remarkably vivid and continuous lightning, are very common round the town of Monghyr; it is believed that the lightning is due to the large quantities of iron ore in the neighbouring hills.

From the middle of May till the breaking of the monsoon the heat is oppressive in the day, and the nights are close and sultry. In normal years the rains commence in June and last till October, the average fall at Monghyr being 7·27 inches in June, 12·32 inches in July, 12·77 inches in August, and 8·67 inches in September. The cold weather sets in during November, when the mornings and evenings are pleasantly cool. In December and January the weather is delightful, being dry and bracing, while the nights are distinctly cold and a fire is necessary. In some years, indeed, the cold is bitter, fires being essential throughout the day, while driving without furs or their equivalent is an unpleasant experience. The sky is, as a rule, cloudless and only a fraction of an inch of rain falls monthly. In February the nights are still cold, but the days become progressively warmer until the hot weather sets in in March.

Statistics of the average rainfall at the different recording stations are given below. It is to be observed that there are considerable variations from year to year above and below these averages, e.g., in 1903 the fall at Monghyr was 29½ inches and in 1899 it was 60¾ inches.

	Begusarai.	Monghyr.	Jamui.	Gogri.	Jamalpur.	Sheikhpura.	Chakai Bamda.	Bakhtiarpur.	Gidhaur.	Kharagpur.
January ...	0·62	0·52	0·64	0·83	0·54	0·49	0·62	0·25	0·59	0·60
February ...	0·79	0·71	0·86	0·97	0·81	0·69	1·11	0·64	1·12	1·09
March ...	0·46	0·42	0·46	0·55	0·50	0·37	0·76	0·55	0·59	0·71
April ...	0·57	0·55	0·43	0·57	0·57	0·30	0·67	0·45	0·66	0·52
May ...	3·22	2·42	1·91	3·06	2·26	1·91	2·58	2·70	2·19	2·08
June ...	6·89	7·27	6·96	8·28	8·69	7·02	8·49	6·27	7·20	8·78
July ...	11·44	12·32	12·81	12·91	12·32	10·38	13·00	10·31	11·16	12·58
August ...	12·12	12·77	11·41	13·26	15·30	11·93	10·65	11·81	9·98	14·19
September ...	7·67	8·67	7·71	9·78	9·08	6·94	9·28	9·15	9·08	8·66
October ...	2·23	2·42	1·95	1·73	1·88	1·72	2·28	1·41	2·08	2·35
November ...	0·38	0·35	0·16	0·34	0·57	0·28	0·24	0·28	0·36	0·50
December ...	0·10	0·12	0·08	0·18	0·01	0·18	0·09	0·27	0·13	0·12
Annual Average ...	45·49	48·54	45·38	52·46	52·53	42·21	49·77	44·09	45·14	52·18

CHAPTER II.

HISTORY.

EARLY HISTORY.

THE territory now included within the district of Monghyr was on the east of the *madhya-desa* or mid-land of the first Aryan settlers; and the alluvial plain which forms the greater portion of it has been ploughed and sown from the earliest times by the same people, swayed by various rulers, who elsewhere have left more tangible records of their dominion. The people, their language, and the prevalence of ideas and customs which have held their ground against successive waves of Muhammadan conquest, still bear strong evidence of Hindu ascendancy. There are, however, no great historic sites associated with ancient Hindu dynasties about which anything can be positively asserted, with the exception of the town of Monghyr itself—and even its history is shrouded by legend and tradition till comparatively recent times. It seems at least certain, however, that its position on the bank of the Ganges, commanding that highway of commerce as well as the land route between the hills and the river, must have led to its occupation at a very early date, and it has been identified with Modagiri, a place mentioned in the *Mahabharata*, which was the capital of a kingdom in Eastern India near Banga and Tamralipta, i.e., Bengal and Tamluk.*

At the dawn of history, the present site of the town was apparently comprised within the old Hindu kingdom of Anga, the capital of which was at Champa near the wodern Bhagalpur, while a portion of the west of the present district was included within the limits of the kingdom of Magadha. Anga was an independent kingdom till the sixth century B.C., and there are traditions of war between it and Magadha. During the lifetime of Buddha it was annexed by Bimbisara, the ambitious ruler of Magadha (cir. 519 B.C.), and it appears never to have regained its independence, the Raja of Anga in the time of Buddha being simply a wealthy nobleman, of whom nothing is known except that he granted a pension to a Brahman.† Thenceforward its history is merged in that of the Magadhan empire.

* A Cunningham, *Ancient Geography of India* (1871), p. 476.

† T. W. Rhys Davids, *Buddhist India* (1903), pp. 23-24.

Hiuen Tsiang's account.

The first historical account of the district as now constituted appears in the Travels of Hiuen Tsiang, who visited portions of it towards the close of the first half of the seventh century A.D. Travelling north-eastwards from the Gaya district, the Chinese pilgrim arrived at a large and populous village to the south of the Ganges, which contained many Brahmanical temples adorned with fine sculptures, as well as a great stupa built on a spot where Buddha had preached for one night. Both the distance and direction mentioned in his account point to the vicinity of Sheikhpura, a supposition which is confirmed by his subsequent easterly route through forests and hill gorges. Leaving Sheikhpura, Hiuen Tsiang travelled for 100 *li*, nearly seventeen miles, to the east, through forests and hills, to a monastery in the village of Lo-in-ni-lo, before which was a stupa erected by Asoka on a spot where Buddha had preached the law for three months. This latter place has been identified by General Cunningham with Rajaona, two miles to the north-west of Lakhisarai, a village which contains several Buddhistic remains. Thence the pilgrim proceeded to the east through " forests and gorges of wild mountains " (presumably the Kharagpur Hills) into the country of I-lan-ha-po-fa-to (*Hiranya-parvata*, the golden mountain), of which he left a detailed account.

" The country," he said, " is regularly cultivated and rich in produce, flowers and fruit being abundant; the climate is agreeable; and the manners of the people are simple and honest. There are ten Buddhist monasteries with about 4,000 priests, and a few Brahmanical temples occupied by various sectaries." The capital, Hiranya-parvata, lay on the southern bank of the Ganges, and close to it stood mount Hiranya, which " belched forth masses of smoke and vapour that obscured the light of the sun and moon ". The position of this hill is determined, from its proximity to the Ganges, and from its bearings and distances from other places mentioned, to be Monghyr; and though no smoke now issues from any peak, the numerous hot springs in the hills within a few miles of the town point to former volcanic action. These hot springs, moreover, are mentioned by Hiuen Tsiang. On the western frontier was another hill with two peaks, where Buddha once stayed in retreat during the three months of the rainy season and overcame the demon Vakula; and to the west were some springs, the water of which was "extremely hot ". This place has been identified by General Cunningham with the hill of

Mahadeva, the hot springs being those of Bhimbandh; but other authorities locate the site at Uren, three miles west of the Kajra railway station.

The kingdom described by Hiuen Tsiang has been held by recognized authorities to have coincided approximately with the south-eastern portion of the district of Monghyr, with its capital at the site of the present town of Monghyr. It was bounded by the Ganges on the north, by forest-clad mountains on the south, probably extending in the latter direction as far as the famous Parasnath Hill, by Magadha on the west, and by Champa on the east. It was probably the king of Champa, a good friend of the Buddhist priests, who had recently deposed its ruler and established two monasteries in its capital.*

The Pala kings.

There is a gap in the history of the district for several centuries after Hiuen Tsiang's visit. In the ninth century A.D. it passed under the rule of the Pala kings, of whose dynasty there is an interesting relic in a copper-plate, known as the Monghyr plate, which was discovered among some ruins at Monghyr about 1780, and translated by Sir Charles Wilkins in 1781. The inscription, which is in Sanskrit, opens with the name of Gopala, who from the context was obviously a pious Buddhist. It mentions his son Dharma Pala, who, according to Dr. Hoernle,† resided in Monghyr about 830 A.D., and records the conquests of Deva Pala, the third of the line. It eulogizes all three as powerful monarchs who conquered almost the whole of India. The immediate purpose of the inscription seems to have been to record the grant by the king last named to a Brahman of certain land in Srinagara (the modern Patna). The place where it was executed was Mudgagiri, i.e., Monghyr, " where is encamped his victorious army, across whose river a bridge of boats is constructed for a road, which is mistaken for a chain of mountains, where immense herds of elephants, like thick black clouds, so darken the face of day, that people think it the season of the rains; whither the princes of the north send so many troops of horse, that the dust of their hoofs spreads darkness on all sides; whither so many mighty chiefs of Jambudwipa resort to pay their respects, that the earth sinks beneath the weight of the feet of their attendants". According to Dr. Rajendra Lala Mitra, Deva Pala reigned between 895 and 919 A.D., and according to Mr. Vincent Smith, between

* S. Beal, *Buddhist Records of the Western World* (1884), Vol. II, pp. 184-191; A. Cunningham, *Ancient Geography of India* (1871), pp. 576-78; Reports Arch. Surv. Ind., Vol. XV, pp. 13-19.

† Hoernle and Stark, *History of India* (1903), p. 72.

853 and 893 A.D., while Professor Kielhorn agrees with Sir A. Cunningham in assigning his reign to about the end of the ninth century A.D.*

Another plate, known as the Bhagalpur plate, by which a similar grant of land was made, was executed at Monghyr by the fifth of the line Narayan Pala, who is called Lord of Anga, i.e., king of Bhagalpur and its neighbourhood, including Monghyr. This plate also refers to Gopala and Deva Pala, while Narayan Pala, the reigning sovereign, is naturally spoken of in the highest terms of praise; but the only notable work of his described in this record is the construction of a bridge of boats across the Ganges near Monghyr. His reign is believed by Dr. Rajendra Lala Mitra to have lasted from 935 to 955 A.D. Like the Monghyr plate, this inscription commemorates a great meeting of the princes and armies subject to the Pala king in his victorious camp at Mudgagiri, and similarly states that his numberless elephants darkened the face of the earth, while the dust from the feet of the horses of "the only king of the north" spread darkness all around.†

From the context it appears that the Pala kings were earnest Buddhists, and the dynasty appears never to have relinquished its allegiance to Buddhism. Owing to their patronage, Bihar, including Monghyr, remained the last refuge of Buddhism in Northern India up to the time of Muhammadan conquest, and numerous Buddhistic remains dating back to the time of the Pala kings are still extant in the district. According to tradition, the last Hindu ruler was Indradyumna [Pala], who was reigning at the time of the Muhammadan conquest; and forts attributed to him are still pointed out in the district.

MUHAMMADAN RULE.

Towards the close of the 12th century, when Bakhtiyar Khilji invaded and conquered Bihar, Monghyr passed under the rule of the Muhammadan conquerors, without apparently offering any resistance. After the conquest, the town of Bihar (now in the Patna district) was made the seat of the Muhammadan governors, and Monghyr seems to have become the second town in South Bihar. It shared the fate of the Province and was attached to Bengal till 1330, when Muhammad Tughlak annexed it to Delhi. An interesting relic of the rule

* Prof. F. Kielhorn, *The Mungir Copper-plate Grant of Devapaladeva*, Indian Antiquary, Vol. XXI, 1892. A translation of the inscription is given by General Cunningham in Reports, Arch. Surv. Ind., Vol. III, 1873 (pp. 114-117).

† Rajendra Lala Mitra, The *Pala and Sena Rajas of Bengal*, J.A.S.B., Vol. XLVII, Part I, 1878.

of the Bengal Sultans still exists at Lakhisarai, in a basalt slab of stone with an inscription bearing a date corresponding to 1297, which mentions Rukn-ud-din Kaikaus (king of Bengal from 1291 to 1302), and a Governor named Firoz Aitagin, who is referred to as a second Alexander and as the Khan of the Khans of the East and of China. From 1397 Monghyr belonged to the kingdom of Jaunpur and continued to form part of it about a hundred years, until the time of Bahlol Lodi, on whose death in 1488 his son Sikandar overran Bihar as far as the frontier of Bengal. At this time, it was chiefly in the hands of Afghan chiefs, but about 1494 the Afghans seem to have submitted to Sultan Husain Shah, king of Bengal, and historians tell us that Prince Danyal, son of Husain Shah, met Sultan Sikandar Lodi of Delhi near Bihar, in the year 1499, when the province was formally acknowledged to belong to Bengal. Stewart, in his *History of Bengal*, gives a different account of this transaction. He states that two noblemen, on the part of Sikandar, met the prince in the town of Barh, and concluded a treaty, the terms of which were that the Emperor should retain Bihar, Tirhut and *Sarkar* Saran, provided he did not invade Bengal.

Prince Danyal appears to have represented his father as Governor of Eastern Bihar. He repaired the fortifications of Monghyr and built, in 1497, the vault over the shrine of Shah Nafah, the Muhammadan patron of the town. This is shown by an inscription put up by Danyal on the eastern wall of the *dargah* or shrine, which lies on an elevated spot, reached by a flight of steps, just within the southern gate of the fort. The *khadims* or custodians of the shrine say that when the fortifications were being repaired, Danyal dreamed that a grave near the wall emitted a smell of musk. The grave was discovered, and the prince built a vault over it, for the tenant of the tomb was clearly a holy man. From this circumstance, the saint is called to this day Shah Nafah, from the Persian *nafah* meaning a pod of musk.

In 1521 Nasrat Shah, one of the eighteen sons of Husain Shah, succeeded his father, and at once took advantage of the troubles of the Emperor of Delhi, who was being pressed by the Mughal invader Babar, to break the treaty which his father had concluded and invade Tirhut. He made his son-in-law, Makhdum Alam, Governor of the conquered territory with his headquarters at Hajipur, and then crossed the Ganges and took possession of the fort and district of Monghyr, which he entrusted to one of his best generals, named Kutub Khan. After this, Monghyr became the headquarters of the Bihar army of the rulers of Gaur, and we learn from the *Memoirs of Babar* that, when Babar invaded Bihar, the prince of Monghyr wrote to him accepting terms of peace for Nasrat Shah after the battle on the Gogra (1529-30). Kutub Khan appears to have remained at Monghyr until the accession of Mahmud Shah, the last independent king of Bengal (1533-4), when Makhdum Alam raised the standard of revolt in concert with the turbulent Afghan chief, Sher Shah. Kutub Khan was directed to advance against the allied rebels, and his defeat by Sher Shah was the first great success of that chief in the struggle which afterwards placed him on the throne of Delhi. Sher Shah next defeated Mahmud Shah, captured the fort of Rohtas, and marching to Monghyr put to the sword the Emperor's noblemen there.* During the subsequent war between him and the Emperor Humayun, when the latter was retreating from Bengal, Monghyr was the scene of a battle between the Afghans and the Emperor, in which Sher Shah captured Dilawar Khan, son of Daulat Khan Lodi, and ancestor of Khan Jahan Lodi, a man of note in the reign of Shah Jahan.

From 1545 we find Monghyr mentioned as in the possession of Miyan Sulaiman, an Afghan of the Kararani tribe, who held South Bihar for Islam Shah, son of Sher Shah. Under Islam Shah's successor, Muhammad Adil Shah, Sulaiman Kararani, desirous of establishing his independence, entered into an alliance with Bahadur Shah, king of Bengal; and when Adil Shah retired before the advancing army of Akbar, Bahadur Shah and Sulaiman attacked him near Surajgarha, west of Monghyr, and defeated and killed him (1557). The battlefield is located at Fatehpur, four miles west of Surajgarha. In 1563, Sulaiman became ruler of Bengal and Bihar, but acknowledged the suzerainty of Akbar. He was succeeded, in the beginning of 1573, by his second son, Daud Shah, who refused to pay tribute to the Mughal Emperor. Next year, therefore, Akbar invaded Bihar and conquered it.

In 1580 the great Bengal military revolt commenced, and Monghyr was for some time the headquarters of Akbar's officers in their expeditions against the rebels. Akbar, fearing to trust his Mughal officers, entrusted the command to the Hindu

* *Riyazu-s-Salatin,* translated by Maulavi Abdus Salam (1904), p. 142.

Raja Todar Mal, who marched southwards without opposition till he reached Monghyr. Here he learned that the rebel army, consisting of 30,000 cavalry, were encamped at Bhagalpur, waiting to give him battle. The Raja took possession of Monghyr, and employed his troops in constructing additional lines, extending to the hills, which, with the fort, completely covered the front of his army. Both armies then remained inactive for several months, but frequent skirmishes took place between their light troops. At length the Raja, by means of his influence among the Hindu zamindars, prevailed upon them no longer to supply the rebels with provisions, promising to pay them ready money for everything brought to his camp. "The combined effects of similarity of religion and ready money payments worked so effectively on the zamindars, that famine shortly found its way into the rebel camp, and compelled the chiefs to separate, in order to obtain food." One body of the rebels made a dash on Patna closely followed by Todar Mal, and the campaign ended with their defeat near Patna. They then fell back on Bengal, and the Province of Bihar was restored to the Mughal Emperor.*

Kharagpur Rajas.

Although the imperial authority was recognized, the semi-independent Rajas of Kharagpur continued for some time to hold their ground in the south of the district. At the time of the conquest of Bihar and Bengal by Akbar (A.D. 1574-5), there were in Bihar three powerful zamindars,—Raja Gajapati of Hajipur, Raja Puran Mal of Gidhaur, and Raja Sangram of Kharagpur. Gajapati was totally ruined by the imperialists, whom he opposed; but Puran Mal and Sangram wisely submitted, and assisted Akbar's generals in the wars of the Afghans. When the great mutiny of Bihar and Bengal broke out, Sangram, though not perhaps very openly, joined the rebels, but submitted again to the Mughals, when Akbar's general Shahbaz Khan marched against him. He was so anxious to avoid coming into open conflict with Akbar, that he handed over to Shahbaz the strong fort of Mahda, a place about six miles north of Kharagpur, where at the present day there are no remains of fortifications. After this, though he never paid respects personally at Court, where his son was detained, apparently as a hostage, he remained submissive till Akbar's death (1605). The accession of Jahangir and the rebellion of Prince Khusru inclined him to make a final attempt to recover his independence; and he collected his forces, which,

* C. Stewart, *History of Bengal* (1847), p. 109.

according to Jahangir's memoirs, consisted of about 4,000 horse and a large army of foot-soldiers. Jahangir Kuli Khan Lala Beg, Governor of Bihar, lost no time in opposing him, and Sangram, whilst defending himself, was killed by a gun-shot (1606).

Sangram's son whom Jahangir calls a favourite of his, was not immediately installed on his father's death, but had to wait till 1615, when, on his conversion to Islam, he was allowed to return to Bihar. Like several other Rajas, he retained, after his conversion, the title of his ancestors, and is known in Muhammadan histories as Raja Rozafzun (i.e., daily growing in power). He remained faithful to the service of the Emperor, and in 1628, when Jahangir died, was a commander of 1,500 foot-soldiers and 700 horse. On Shah Jahan's accession, Raja Rozafzun entered active service. He accompanied Mahabat Khan to Kabul, in the war with Nazr Muhammad Khan, king of Balkh, and served later in the expedition against Jhujhar Singh Bundela. In the sixth year of Shah Jahan's reign he took part under Prince Shuja in the siege of Parendah, and was promoted in 1634-5 to a command of 2,000 foot and 1,000 horse. He died soon afterwards in the same year and was succeeded by his son Raja Bihruz, who served in the siege of Kandahar and was commander of 700 foot and 700 horse under Shah Jahan. In the beginning of Aurangzeb's reign he assisted the Emperor against Shah Shuja, and also in the conquest of Palamau in 1661, dying four years later in 1665.* The subsequent history of the family will be given in the article on Kharagpur in Chapter XIV.

SHAH SHUJA.

Shah Shuja mentioned above was the second son of Shah Jahan and Governor of Bengal. On hearing of the dangerous illness of his father in 1657, he raised the standard of revolt and claimed the imperial throne. His capital was at Rajmahal, but Monghyr formed the centre from which he directed his preparations, and to which he retired next year after his defeat at Bahadurpur, near Benares, by Sulaiman, the son of his brother Dara Shukoh. The lines of Monghyr held out against the victor, till he was summoned back to Agra to assist his father against Aurangzeb. Subsequently, when Aurangzeb had seized the throne, Shah Shuja again advanced his claims, and marched northwards with a large army until he met with a decisive defeat at Kudha. Once again he fled precipitately back to Monghyr, where he was joined by a number of his dispersed

* Proceedings of the Asiatic Society of Bengal for 1870, pp. 306-307.

followers and by some reinforcements from Bengal. He strengthened the fortifications of Monghyr by deepening the moat, and threw up entrenchments from the fort to the hills, which formed a complete barrier against the enemy's advance on that side of the river.

His position, however, was rendered untenable by the strategy of his pursuers. Muhammad, the son of Aurangzeb, who had followed hard after him as far as Patna, was joined there by Mir Jumla and the main body of the army. The latter was now detached with a picked force of 12,000 cavalry through the passes of Sherghati to the south of the Gaya district, while Muhammad, with the remainder of the royal army, proceeded by easy marches towards Monghyr, and encamped within a few miles of the fort. For several days skirmishing parties advanced from the camp, and made a show as if they intended to assault the entrenchments. By these means the attention of Shuja was occupied, till his couriers brought him information that Mir Jumla had entered the country in his rear, and was advancing against his capital. His position being thus turned, Shah Shuja immediately withdrew all his troops and retired on Rajmahal, and then, finding that position also untenable, fled to Bengal.

MUHAMMADAN GOVERNORS.

There is little other information regarding Monghyr at this time. There are however a few references to the district in the *Ain-i-Akbari*, according to which *Sarkar* Monghyr consisted of 31 *mahals* or *parganas*, paying a revenue of 109,625,981 *dams* (40 *dams* being equal to one Akbarshahi rupee), and furnished 2,150 horse and 50,000 foot-soldiers. These numbers, however, are perhaps nominal rather than real, for south of Monghyr the country was mostly in the hands of the Rajas of Kharagpur. *Pargana* Monghyr itself was assessed at 808,907½ *dams*. Monghyr was also for some time the residence of Raja Man Singh, who reconquered Bengal and Orissa; and it was here that a pious Musalman, named Shah Daulat, whom Man Singh favoured, tried his best to convert him to Islam. During the reign of Jahangir, Kasim Khan, brother of Ali-ud-din Islam Khan, was in charge of *Sarkar* Monghyr, and on the death of his brother, became Governor of Bengal. Two *jagirdars* of Monghyr are also mentioned, Sardar Khan and Hassan Ali Khan (1619). In the first year of Shah Jahan's reign (1628) Saiyad Muhammad Mukhtar Khan was appointed *tayuldar* of Monghyr. He distinguished himself in a campaign against the Raja of Dumraon about 1637. Another *tayuldar* was Mahaldar Khan.

The historians of Aurangzeb's reign mention only one other event in connection with Monghyr, viz., the death and burial at Monghyr of the poet Mulla Muhammad Saiyad, who wrote under the *nom-de-plume* of Ashraf. He was the son of Mulla Muhammad Saleh of Mazandaran, near the Caspian Sea, and stood in high favour with prince Azim-us-Shan, Aurangzeb's grandson, who was Governor of Bihar. He had also been for a long time the teacher of Zebunnissa Begam, Aurangzeb's daughter, herself a poetess of repute. In 1704, while on his way from Bengal to Mecca, the poet died at Monghyr, where his tomb is still pointed out.

LAST DAYS OF MUHAMMADAN RULES.

In the seventeenth century Monghyr appears to have been a well fortified town with fine buildings, and Nicolas Graaf*, a Dutch physician who visited it in the beginning of that century, was struck with admiration at the sight of its white walls, towers and minarets. In 1745, however, when Mustapha Khan, a rebellious general of Ali Vardi Khan, advanced against it in his march northwards, the fort was " a ruinous fortification which though quite neglected had some renown ". Mustapha Khan found it sufficient to detach a brigade under his nephew Abdul Rasul Khan for its capture. " The Governor and his little garrison put themselves upon their defence; but as the fortifications did not seem to deserve much ceremony in that officer's opinion, he alighted and putting himself at the head of his men, mounted to the assault. In an instant the besiegers got upon the wall, and seized the fort; but their leader was killed outright by a stone that fell upon him, on its being loosened by one of the garrison who stood above. The place was not such an acquisition as could compensate the loss of so valiant an officer." † Mustapha Khan, following the custom of those days, had music played to celebrate his success, took some guns and ammunition from the fort, and after a halt of three days marched off towards Patna.

Maratha raid.

This was not the first nor the last of the armies which passed through Monghyr during the eighteenth century. Only two years before, Balaji Rao, Peshwa of Poona, had marched

---

* On Graaf's imprisonment at Monghyr, and his account of the fort, see Mr. C. E. A. W. Oldham's article, *An old description of the Monghyr Fort*, in *Bengal: Past and Present*, Vol. XXVII, part II.

† *Sair-ul-Mutakharin*, Raymond's translation.

through it at the head of a large force when on his way to relieve Ali Vardi Khan and drive Raghuji Bhonsla from Bengal. On this occasion, we are told, the land and its inhabitants suffered exceedingly from the passage of the Maratha troops, large numbers flying in terror to the other side of the Ganges.

Eyre Coote's expedition.

In 1757, Eyre Coote came to Monghyr at the head of a British force on his way up the Ganges in pursuit of Jean Law, the French adventurer and partisan of Siraj-ud-daula, who was flying northwards after the battle of Plassey. Major Coote reached Monghyr late at night on the 20th July and requisitioned a number of boats, which the *Diwan* or Governor of the place supplied. He was not allowed, however, to enter the fort, for when he approached the walls, he received a significant hint, the garrison lining the ramparts with their matches and port-fires lighted. Coote, accordingly, was obliged to resume his march without making any further attempt to examine the fort.

Caillaud's campaign.

Nearly three years later, in the spring of 1760, the army of the emperor Shah Alam marched through the south of the district, pursued by Major Caillaud and Miran. In this, his second invasion of Bihar, the emperor had been defeated by Caillaud and Miran at Sirpur on the 22nd of February. The emperor withdrew to Bihar town, while the victorious army remained between Barh and Patna. When on the 29th of February the allies began to follow up the army of the emperor, marching on Bihar through Chandi, Shah Alam doubled in a northerly direction to the Ganges, with the intention of capturing Murshidabad in Bengal. Caillaud pursued him; and the emperor, fearing that if he attempted to reach Bengal by the Ganges road he would be overtaken and intercepted at Sakrigali, turned towards the south-east at Lakhisarai, and made for Bengal through Malepur, Chakai and Deogarh, through hills and jungles never before traversed by any army, little known to either his own or the Nawab's troops, and to the English a perfect *terra incognita*. Major Caillaud, who dared not leave the young Nawab unsupported, was compelled to follow the same course, and disembarking his troops once more, joined Miran and commenced the pursuit. The route was most difficult, through thick jungle of the wildest description, across unbridged streams, over hills and valleys, and through difficult and unknown passes. For days together a road had to be made for the field pieces, and the troops suffered extreme fatigue and privations, which they bore without a murmur;

their zeal and exertions, coupled with the energy and resource of the commander and his officers, surmounted all obstacles and enabled them to keep so closely on the track of the enemy, that on the 22nd March, having passed through one of the most difficult defiles, they arrived on the ground which the Emperor had quitted only two days before.*

Stables' campaign.

At the end of the same year another small campaign was fought out in the Kharagpur Hills. John Stables, then an Ensign and subsequently a Member of Council, had been left by Caillaud at Monghyr, and was now directed to attack the Kharagpur Raja, who had openly defied the authority of the new Nawab, Kasim Ali Khan. His detachment consisted of three companies of sepoys; a party of about 50 or 60 pseudo-Europeans in the Nawab's service, composed of renegade deserters, Armenians, Portuguese, and "Topasses"; and two troops of Mughal horse, making a total of about 550 men. Hearing of his intentions, the Raja sent a force of 2,000 men, which encamped about three miles from Monghyr. At one o'clock next morning Stables marched out, hoping to surprise the enemy's camp, but the alarm had been given, and he found them strongly posted under cover of an old entrenchment. He did not hesitate, however, to attack, and finally succeeded in forcing the camp at the point of the bayonet. By this time day was beginning to break, and he determined to follow up his blow by advancing at once upon the Raja's position at Kharagpur, a distance of about sixteen miles. He accordingly sent forward his cavalry in pursuit of the fugitives, and advanced more steadily with his infantry. After about four hours' march, they arrived within three miles of Kharagpur, where they found the Raja had drawn up his whole force to oppose them. Attaining the enemy's flank, Stables soon put them to flight, and followed them to Kharagpur, when they rallied and made a desperate stand in the Raja's palace. Here also they were again completely defeated, and the British troops having dispersed them, set fire to the town and palace, and reduced the whole to ashes. The following morning they returned to Monghyr, having in one day marched upwards of twenty miles, been engaged three times, and completely destroyed the force of an enemy ten-fold their own number,—a proceeding that speedily resulted in the submission of the Raja and the restoration of tranquillity in the district.†

* A. Broome, *History of the Rise and Progress of the Bengal Army*, (1850), Vol. I, pp. 288-289.
† *Ibid*, p. 320.

KASIM ALI KHAN'S RULE.

Monghyr again came into prominence in 1761, when Kasim Ali Khan made it his capital instead of Murshidabad, from which he removed his treasure, his elephants and horses, and even the gold and silver decorations of the Imambara. He built himself a palace with a breast-work before it for thirty guns, and had the fortifications strengthened; while his favourite general Gurghin (Gregory) Khan, an Armenian who had formerly been a cloth-merchant at Ispahan, reorganized the army, had it drilled and equipped after the English model, and established an arsenal: the manufacture of fire-arms, which still is carried on at Monghyr, is said to date back to this period. A picturesque account of the Court of the Nawab at Monghyr has been left in the *Sair-ul-Mutakharin*. Two days a week he sat in a public hall of audience and personally dispensed justice, listening patiently to the complaints and grievances of everyone, however humble his position, nor dared anyone in his court touch a bribe or advocate an unjust claim. The poor, defenceless landholders, who had been dispossessed of their villages and hereditary estates by rapacious zamindars, now found that the day of redress was come; for their claims were heard, and, if proved, mace-bearers were sent to see the oppressed reinstated in their estates and the defenceless righted. The Nawab, indeed, was a terror both to his enemies and to wrong-doers. He also honoured learning and the learned, and welcomed scholars and savants to his court, so that friends and foes alike respected him. The author of the *Sair-ul-Mutakharin* admits, it is true, that his temper had been soured and rendered suspicious by perpetual treasons, duplicities, and infidelities, and that he was "ever prone to confiscation of properties, confinement of persons, and effusion of blood. Nevertheless, it was acknowledged on all hands that he had admirable qualifications that balanced his bad ones," and made him "an incomparable man, indeed, and the most extraordinary prince of his age".

Soon after he had established himself at Monghyr, Mir Kasim Ali came into collision with the English. The first quarrel appears to have been caused by the tactless conduct of Mr. Ellis, who was in charge of the English factory at Patna. A vague report having reached the latter that two English deserters were concealed at Monghyr, he sent a company of sepoys under a sergeant to search the fort. They were refused admission, and this was construed by Mr. Ellis into an act of hostility, while the Nawab complained of the insult offered to his authority. A long dispute followed, which was finally compromised by Lieutenant Ironside, the Town Major of Calcutta, being specially sent to search the fort with the permission of the Nawab. No deserters were found, the only European in the place being an old French invalid. About the same time (April 1762) Warren Hastings was sent up from Calcutta to arrange terms between the Nawab and Ellis, and was favourably received by the former in his camp at Sasaram. Ellis, however, refused to meet Warren Hastings at Patna and stayed in his house at Singia, fifteen miles away, saying that he could not be expected to pay him the compliment of travelling such a distance in the hot weather!

Trade disputes.

A more serious dispute now arose. The East India Company had long enjoyed exemption from the heavy transit dues levied on inland trade. After the battle of Plassey the European servants of the Company began to trade extensively on their own account, and to claim a similar exemption for all goods passing under the Company's flag and covered by a *dastak* or certificate signed by the Governor or one of the factory agents. The English had thus a great advantage over their rivals and most of the trade passed into their hands or was carried on under their name. Great abuses followed, for the English in some cases lent their names to Indians for a consideration, and the latter used the same *dastaks* over and over again or forged them. If the Nawab's collectors attempted to check these malpractices, they were seized by the nearest English agent, confined and punished. Every one who could hire a *dastak* or fly the Company's flag, did so; and matters were made worse by the rascality of their Bengali employés, who had previously been walking about in rags at Calcutta, but now assumed the dress of sepoys, and flogged and imprisoned those who refused to buy from or sell to them at their own price. On his way to Patna in 1762, Warren Hastings say that every boat he met on the river bore the Company's flag, and became aware of the oppression of the people by the *gumashtas* and the Company's servants. "Most of the petty towns and *sarais*," he wrote, "were deserted at our approach, and the shops shut up from the apprehension of the same treatment from us."

Vansittart's visit.

Mir Kasim bitterly complained that his chief source of revenue had been taken from him and that his authority was completely disregarded. Eventually, in December 1762,

Mr. Vansittart, the Governor, left Calcutta in order to try and conclude a settlement that would be satisfactory to both parties. He found the Nawab at Monghyr smarting under the injuries and insults he had received; but at length it was agreed that the servants of the Company should be allowed to carry on the inland private trade, on payment of a fixed duty of nine per cent. on all goods—a rate much below that paid by other merchants—and that, to prevent abuses, *dastaks* should be signed by the Agents of the factories through whose circles the goods passed, and also countersigned by the Nawab's collectors. Kasim Ali Khan agreed to these terms very unwillingly.

A picturesque account of Vansittart's visit is given by Ghulam Husain Khan the author of the *Sair-ul-Mutakharin*, who held a *jagir* in the district. The Nawab advanced six miles to meet Vansittart and assigned for his residence a house which Gurghin Khan had erected on "the hill of Sitakund". He gave him a public reception in the hall of audience, where he seated him on his own *masnad*, entertained him with a *nach*, paraded troops in his honour, and, last but not least, "complimented him with a variety of curious and costly presents suitable to the occasion". It is said, indeed, that Vansittart received five lakhs of rupees, of which three lakhs were paid to him at Monghyr, and that he allowed two ladies who accompanied him to go into the zanana of Mir Kasim Ali and receive presents of jewelry.

The Company's claims.

After staying a week at Monghyr, Vansittart returned to Calcutta in January 1763, only to find that the agreement concluded with the Nawab was repudiated. The Nawab, however, had sent copies of the Governor's agreement to all his officers directing that it should be acted upon at once. The consequence was that the English goods then in transit were stopped and duty claimed upon them. The Council were indignant, and at once declared that all goods, except salt, covered by an English *dastak* should pass free of duty. The Nawab, on the other hand, protested at this breach of faith and passed orders abolishing all transit duties and throwing open the whole inland trade. The English regarded this as an act of hostility, and preparations were made to resist his measures and declare war against him, if he persisted. But, as a last resource, it was agreed that a deputation, headed by Messrs. Amyatt and Hay, should be sent to Monghyr to endeavour to arrange terms with the Nawab. Mr. Ellis was informed of this measure, and warned to commit no act of aggression, even should the mission be unsuccessful, until Amyatt and Hay were well out of the Nawab's power.

Negotiations with the Nawab.

The party reached Monghyr on the 14th May 1763 and opened up negotiations, but it was soon seen that they were unwelcome. At the first interview the Nawab, offended at the rough and overbearing manner in which he was addressed by the English linguist, refused to speak to him. At subsequent interviews he himself never failed to do something to offend the English; and he steadily refused to come to terms. The envoys were kept under strict surveillance, and on one occasion, when some of the party wished to ride out from Monghyr, they found their way barred by the Nawab's soldiers with lighted matches ready to fire. Unfortunately too just at this time, an English cargo boat from Calcutta was detained at Monghyr, and five hundred muskets intended for the factory at Patna were found hidden under the cargo. The Nawab insisted that the English intended to seize the fort and city of Patna, and demanded that their detachment at the factory should be withdrawn to Monghyr, where they would be checked by his own troops; otherwise, he would declare war. In the meantime, he permitted Mr. Amyatt and others of the party to leave for Calcutta, but detained Mr. Hay and Mr. Gulston as hostages for the safety of his officers who had been arrested by the English.

Outbreak of war.

A final rupture was now imminent, and this was brought about by the precipitate action of Mr. Ellis, who, in the belief that war was in any case inevitable, seized the city of Patna, on hearing that a detachment was advancing from Monghyr to reinforce the Nawab's garrison. The Nawab retaliated promptly, reinforcements were hurried up, and the fort quickly recaptured. The news of this success gave Kasim Ali the keenest delight. Though it was the middle of the night, he immediately ordered music to strike and awaken the whole town of Monghyr. At daybreak, the doors of the public hall were thrown open, and every one hastened to offer him congratulations. He now proclaimed the outbreak of war between himself and the English, and directed his officers to put the latter to the sword wherever found. In pursuance of this general order, Mr. Amyatt was killed at Murshidabad, and the factory at Cossimbazar was stormed. The survivors surrendered, and were sent to Monghyr to join their unfortunate companions from Patna.

The Nawab's prisoners.

A British force under Major Adams quickly advanced against the Nawab, and defeated his troops at Suti. On hearing of this defeat, he sent his Begams and children to the fort at Rohtas and set out himself, accompanied by Gurghin Khan, to join his army that was now concentrated on the banks of the Udhua Nullah near Rajmahal. Before leaving Monghyr, however, he put to death a number of his prisoners, including Raja Ram Narayan, till lately Deputy Governor of Bihar, who was thrown into the river below the fort with a pitcher filled with sand tied to his neck. Gurghin Khan, not satisfied with this butchery, also urged the Nawab to kill his English prisoners, but this the Nawab refused to do.

Jagat Set Mahtab Rai and Sarup Chand, two rich bankers of Murshidabad, who had been brought from that place by Mir Kasim Ali, as they were believed to favour the British cause, also appear to have escaped; though tradition says that they were drowned at the same time, and that a servant of Jagat Set, Chuni, begged in vain that he might be drowned with his master, and thereupon flung himself into the water after him. This picturesque tradition appears, however, to be untrue, for not only does the author of the *Sair-ul-Mutakharin* say that they were hacked to pieces at Barh, but Major Grant, who was one of the avenging force which hurried up under Major Adams, states that they found the bodies buried in one of the rooms of a house at that place. The tradition is, however, an old one, which must have sprung up soon after their death; for the translator of the *Sair-ul-Mutakharin* (Raymond *alias* Haji Mustapha) wrote about 1789 :—" Out of 10,000 boatmen who pass every year by a certain tower of the castle of Monghyr, there is not a man but will point out the spot where the two Jagat Sets were drowned, nor is there an old woman at Monghyr but would repeat the speech of the heroical Chuni to his master's executioners."

The Nawab's flight.

Before the Nawab could join his army at Udhua Nullah, he heard of a second decisive defeat that it had sustained, and he thereupon returned to Monghyr. He only stayed there two or three days, and then fell back on Patna, taking with him Mr. Hay, Mr. Ellis, Mr. Lushington and other English prisoners. Before leaving, it is said, one of his noblemen Ibrahim Ali Khan (brother of the ancestor of the Husainabad family in this district) urged him to release the English prisoners, or at least send the women down the Ganges to Major

Adams; but the Nawab simply referred him to his evil genius, Gurghin Khan, who put him off by saying that it would be impossible to find boats for so many.

On the way Kasim Ali Khan halted on the banks of the Rahua Nullah, a small stream near Lakhisarai. Here Gurghin Khan met his death, being cut down by some troopers who demanded their arrears of pay. A scene of wild confusion followed. Makar, another Armenian general, seeing the murderers were making their escape beyond the reach of muskets, fired off some guns loaded with grape. The army thought that the English were upon them and fled in terror, Mir Kasim himself trying to escape on an elephant. They rushed to the bridge of boats over the Rahua, which was densely crowded with fugitives, carts and elephants, all trying to cut their way through; and as soon as they had crossed over, made ready to resist their imaginary pursuers. The uproar did not subside till midnight, and the alarm having been proved to be a false one, Mir Kasim marched on next day to Patna.

Capture of Monghyr.

In the meantime, the British army continued to advance upon Monghyr, which Mir Kasim had placed under the command of Arab Ali Khan, a creature of Gurghin Khan. As they approached the place, a strong detachment was sent forward to invest it and commence approaches. On the 1st of October 1763 the main body arrived, and the batteries which had been thrown up were immediately opened. A heavy fire was maintained all that and the following day, when the breach was reported practicable, and arrangements were made for an assault. But in the evening the Governor capitulated, and surrendered himself and his garrison prisoners of war. The English at once set to work to repair the breaches and improve the defences. The sick and wounded were disembarked from the boats and brought from Rajmahal, and a comfortable hospital established. A depôt of stores was also formed, and a detachment of sepoys was left as a garrison under the command of Captain John White, who was further directed to raise locally another battalion of sepoys. The news of the capture of Monghyr infuriated the Nawab, who as soon as he heard of it, gave orders that his English prisoners at Patna should be put to death. This order was carried out by the renegade Samru, and resulted in what is known as the Massacre of Patna.

Three years later (in 1766) there was a mutiny of the European officers of the Bengal army in consequence of the orders regarding the reduction of *bhatta*. The meaning of this term may be briefly explained. The officers received fixed pay, but when they took the field, were allowed an extra monthly sum to cover their increased expenses. This allowance was known as *bhatta*, and half of it was granted when they were garrisoned at a detached station but not actually in the field. After the battle of Plassey, Mir Jafar Khan had granted an extra allowance called "double *bhatta*" and this had been continued during the Nawabship of Mir Kasim. In pursuance of a policy of rigorous retrenchment, the Directors of the Company now passed orders that this allowance should be abolished, except in the case of the Brigade stationed at Allahabad, but conceded the grant of half *bhatta* to the troops stationed at Patna and Monghyr. The officers bitterly resented this curtailment of their emoluments and determined to resign their commissions. On the 1st May 1766 a letter to this effect was signed by the officers of the First Brigade, which was then stationed at Monghyr under Sir Robert Fletcher, who transmitted it to Lord Clive at Murshidabad.

Clive at once determined to proceed to Monghyr in person by forced marches, and in the meantime sent forward some officers to deal with the situation as well as they could. His confidence in them was not misplaced. Arriving at Monghyr late at night on the 12th May, they heard, much to their surprise, drums beating and other signs of disturbance, and on going to Sir Robert Fletcher's quarters saw half the European regiment drinking, singing and beating drums. Next morning two of them went to Kharagpur, where the sepoy troops were encamped, and returned with two battalions to Monghyr. On the 14th May the European battalion broke out in open mutiny; but this proved abortive owing to the prompt measures taken by Captain Smith. Expecting such a mutiny, he had already determined to seize the saluting battery, which being situated upon a hillock completely commanded the barracks, and from its situation, and the guns mounted upon it, was a position of importance.*

Captain Smith marched the sepoys to the back of the hill without being observed, and making a sudden rush, gained

* This hillock is now known as the Karnachaura hill. See the article on Monghyr in Chapter XIV.

possession of it. The European battalion had got under arms, and were preparing to leave the fort and follow their officers, and the artillery were about to do the same. But the unexpected appearance of the firm line of sepoys, with their bayonets fixed and arms loaded, threw them into confusion, of which Captain Smith immediately took advantage, warning them that if they did not retire peaceably to their barracks he would fire upon them at once. Sir Robert Fletcher, who came up at this juncture, also harangued the troops, and distributed money amongst them. They now became quiet, and said that they had expected he would have placed himself at their head; otherwise they would not have thought of turning out. If such was not the case, their officers might go where they pleased, but they would stay, and live or die with him alone. While Sir Robert was on the parade, several of the officers who had resigned came up, hearing that the men had mutinied, and offered their services to assist him. This offer he declined and ordered them all to leave the garrison within two hours, under pain of being sent off in charge of a guard. In the evening, they had all left the fort except three, of whom two were confined to their beds by sickness.

Clive's visit.

Next day (15th May) Lord Clive reached Monghyr and held a parade of the troops. He explained the circumstances under which the *bhatta* had been withdrawn, applauded the loyal conduct of the sepoys, and pointing out that the combination of the officers was an act of gross insubordination and positive mutiny, declared that the ring-leaders would suffer the severest penalties of martial law, and the inferior offenders be sent to England by the first available ships. After this address the Brigade gave three hearty cheers and marched off quietly to the barracks and lines. The resigned officers had all encamped at a short distance from Monghyr, intending to wait for their comrades of the other Brigades; but Lord Clive ordered them all to proceed to Calcutta forthwith and sent out a detachment of sepoys to take them down by force, if necessary. This measure had the desired effect, and the detachment soon returned with a report that the whole were on their way in small parties of three and four, some marching, and others proceeding by water. Those officers, chiefly subalterns, who were at the outposts, and whose commissions had not been sent in, owing to their absence, were now directed to proceed to Monghyr, where, warned by the example of the others, they exhibited no inclination to resign, but on the contrary, readily undertook the duties of the station, until they were relieved by officers sent up from Calcutta.

EARLY BRITISH ADMINISTRATION.

The threatened mutiny was thus quickly quelled; but it is clear that the danger was very real. One officer, indeed, had proposed that he and the other mutinous officers should throw dice to decide who should take Lord Clive's life, and that when he held a review, they should carry muskets instead of swords in order to enable them to carry out their purpose. It remains to note that Sir Robert Fletcher was subsequently tried for fomenting and encouraging the mutiny and was cashiered.*

THE SEPOY MUTINY.

The subsequent history of the district is uneventful. With the extension of the British dominions the town of Monghyr ceased to be an important frontier post. Although there was an arsenal, no regular garrison was kept up and no attempt was made to bring the fortifications up to date. It was, however, noted for its fine situation and pure air, and was used as a sanitorium for the British troops. So great, indeed, was its reputation as a health resort that the journey up the Ganges, followed by a stay there, was regarded as good as a sea voyage. We find that a trip to Monghyr was prescribed for the wife of Warren Hastings when she was in ill-health; while in 1781, when Warren Hastings was on his way to meet Chait Singh at Benares, he left her here assured that she would be in the surroundings most likely to be of benefit to her.† In the early part of the nineteenth century, the place was degraded to a lunatic asylum for sepoys, a depôt for army clothing and an invalid station for British soldiers, Monghyr being, according to Bishop Heber, generally chosen by the more respectable of the latter, while the reprobates preferred Murshidabad.

During the Mutiny of 1857, the tranquillity of the district was not broken, largely owing to the prompt measures taken by the Commissioner, Mr. Yule. On the outbreak of the troops at Dinapore there was a panic in the town, but Mr. Yule, realizing the importance of its position on the Ganges and the likelihood that the native troops stationed in the Division would rise, detained 150 men of the 5th Fusiliers, who happened to be passing up the Ganges, and sent a hundred of them to

* A. Broome, *Rise and Progress of the Bengal Army*, pp. 561—617.

† S. C. Grier, *The Letters of Warren Hastings to his Wife* (1905), p. 123.

Bhagalpur and the remainder to Monghyr. This prudent measure secured the important highway of the Ganges and enabled Outram to go up country unmolested. Eventually, it was found possible to withdraw the detachment from Monghyr by the end of 1857, and the district remained quiet. There was, it is true, a certain amount of crime, but this was attributed to the scarcity which prevailed at the time and to the distress caused by the stoppage of railway and other works during the disturbances.

The existence of Monghyr as a separate executive centre is believed to date from the year 1812. It appears from a letter dated the 15th July of that year, that Mr. Ewing was appointed to have charge of the Monghyr Criminal Court, called the Court of the Joint-Magistrate of Monghyr, and that he was subordinate to the Magistrate of Bhagalpur, bearing to him somewhat the relation of a modern Subdivisional Officer. The language of the above letter and subsequent correspondence seem to show that the appointment of Mr. Ewing was the first step taken; but the original orders directing the formation of the new jurisdiction cannot be traced in the records of either Bhagalpur or Monghyr. A letter also from Mr. Dowdeswell, Secretary to Government, dated the 22nd October 1811, proves that at that time no magisterial authority existed in Monghyr except that of the Magistrate of Bhagalpur, to whom it is addressed. "I am directed," it runs, " to acquaint you that His Excellency the Vice-President in Council considers it of importance that you should revert to the practice which formerly existed, of holding the *kachahri* during a part of the year at Monghyr, and that he desires that you will make the necessary arrangements for that purpose." The extent of the Monghyr jurisdiction is not mentioned in the local records till September 1814, when it is stated to comprise five thanas or police divisions, viz., Monghyr, Tarapur, Surajgarha, Mallepur, and Gogri.

FORMATION OF TH DISTRICT.

No change seems to have been made in the powers or jurisdiction of the Monghyr Court till 1832, when it was determined to erect it into a revenue-receiving centre under the name of a Deputy Collectorship, the new office being conferred on the then Joint-Magistrate. From that time this officer, although he did not obtain the title, exercised most of the powers of a full Magistrate-Collector, and from the first corresponded directly with the chief executive and revenue authorities, and not through the Collector of Bhagalpur, whose deputy he nominally was. In order to form the new revenue district, *parganas* Surajgarha, Monghyr, Chandanbhuka, Kajra, Pharkiya, Abhaipur and Gidhaur were transferred from the district of Bhagalpur; *parganas* Amarthu, Roh (in part), Narhat (in part), Maldah, Bihar (in part)and Samai (in part) from the district of Bihar; and Malki, Ballia, Masjidpur, Akbarpur-Rani, Bhusari, Badaphusari, Naipur, Imadpur, Kabkhand and Utarkhand from the district of Tirhut.

Since that time the district has gradually grown to its present dimensions with various changes of jurisdiction. In 1834 *pargana* Chakai was transferred from the district of Ramgarh, and in 1839 *pargana* Bisthazari from the district of Patna. Numerous minor changes followed, but the greatest change was effected in June 1874, when *parganas* Sakhrabadi, Darara, Singhaul, Kharagpur, and Parbatpara were transferred to Monghyr from Bhagalpur together with *tappas* Lodwah and Simrawan and 281 villages from *parganas* Sahrui and Lakhanpur, comprising in all an area of 613·62 square miles. The subdivision of Jamui was formed in 1864, at first with headquarters at Sikandra, but in 1869 they were transferred to Jamui. The Begusarai subdivision was established in 1870, the headquarters of that subdivision (formerly known as the Ballia subdivision) being fixed at Begusarai. In recent years the most important administrative change has been the transfer in 1904-5 of the Sheikhpura thana from the Jamui to the Monghyr subdivision.

ARCHÆOLOGY.

The district contains several remains of antiquarian interest. In addition to the great fort at Monghyr, there are the ruins of other forts at Indpe (near Jamui), Naulakhagarh (near Khaira), Chakai, Jaimangalgarh in the Kabar Tal, and Naula in the Begusarai subdivision. Buddhist remains are to be found at Rajaona and Hasanganj near Lakhisarai and at Uren near Kajra. There is an inscription of about the tenth century at Kashtharani Ghat and another referring to the Bengal Sultan Rukn-ud-din Kaikaus (1297 A.D.) at Lakhisarai. The oldest extant building of the Muhammadan period is the *dargah* of Shah Nafah, built in 1497-8 by Prince Danyal, son of Ala-ud-din Husain, king of Bengal.

CHAPTER III.

THE PEOPLE.

THE first census was taken in 1872, and the result was to show for the district as now constituted a population of 1,814,638. During the next nine years there was an increase of 8·5 per cent., the number returned at the census of 1881 being 1,969,950. Part of the increase may have been due to more accurate enumeration; the advance in the next decade was only 3·3 per cent., and the population in 1891 was 2,036,021. At the census of 1901 it was found that the population had increased to 2,068,804, or by 1·6 per cent. A slightly higher rate of increase was shown in 1911, when the population was 2,135,000. In 1921 the population was 2,029,965, showing a decrease of 4·92 per cent. The following table shows the population by thanas, with the percentage of variation since 1901. CHANGES IN POPULATION.

THANA.	POPULATION 1921.	PERCENTAGE OF VARIATION.	
		1911—1921.	1901—1911.
Gogri	412,798	− 2·38	+ 6·00
Monghyr	159,591	+ 3·33	+18·78
Jamalpur	24,827	+20·95	+ 8·27
Surajgarha	74,213	− 0·49	−12·83
Lakhisarai	119,537	− 7·86	+ 9·31
Kharagpur	139,612	− 6·36	− 5·52
Shaikhpura	142,493	− 2·02	+ 2·21
Teghra	215,550	- 10·83	− 1·36
Begusarai	370,649	−12·01	+ 2·93
Sikandra	101,446	− 5·56	− 0·66
Jamui	135,378	− 5·52	+20·17
Chakai	133,871	− 1·46	− 7·97

The following remarks on these figures are quoted from Mr. P. C. Tallents' Report on the census of 1921.

The variation in the population of Monghyr has throughout followed a very similar course to that of the adjoining district of Bhagalpur which it closely resembles in physical characteristics. After the great increase recorded at the second census, Monghyr settled down to a steady growth which has only been checked in the last decade. In the ten years ending in 1891 the increase was greatest north of the Ganges, the southern half of the district having suffered from fever and cholera and a large number of persons especially males having emigrated from this area, from which Calcutta and the coalfields are readily accessible. The census of 1901 was affected by the plague which had recently broken out; the mortality was not serious but the people were terrified at the new and strange disease and left their homes in large numbers. The part north of the Ganges was still immune and the population of the Begusarai subdivision again showed the largest increase. The sadr subdivision was stationary, the variations in the thanas corresponding with the displacement of the population at the time of the plague panic; the Jamui subdivision was stationary also in spite of spreading cultivation in the hill tracts. The period 1901 to 1911 was a prosperous and healthy one in spite of epidemics of plague and fever, and the natural increase of the population was marked. On this occasion the Begusarai subdivision, into which plague had now penetrated, had the smallest increase to show. The sadr subdivision showed a general increase in all thanas except Kharagpur and Surajgarha : the population of the former of these had been swollen by plague refugees in 1901, and the latter was depopulated in 1911 by an outbreak of plague at census time and by an exodus of labourers to cut the crops in Sheikhpura. The increase in the Jamui subdivision was ascribed to the increase of cultivation in Jamui thana.

The year 1911 was unhealthy as elsewhere with higher mortality than usual from cholera, plague and fever; but the number of births was well in excess of the number of deaths in this and every other year of the ten except 1917, 1918 and 1919. In this district also the plague seems to be losing its force, for two-thirds of the total mortality from plague occurred in the first half of the decade. Outbreaks of cholera occurred at intervals; more than five thousand deaths were ascribed to it in 1911, 1915, 1917 and 1918, the last being the worst attack. The deaths from fever increased steadily from 1914 to 1918, the year of the influenza epidemic, and then sank to normal again in 1920. Generally speaking the health of the district was good down to 1916. In that year the Burh Gandak was in flood in the north of the district and damage was done to property and crops, but the public health suffered little. In 1917 cholera, plague and fever brought the death-rate up to 41·6 per mille against an average of 31·8 for the preceding five years. There was also a heavy flood in the Ganges : this damaged the autumn and winter crops and reduced the stocks in hand which were further depleted by exports in response to high prices in the early months of 1918. The outbreak of cholera in 1918 was of exceptional violence, and caused 16,000 deaths. In August there was heavy rain which damaged the *bhadai* crops and in September the rains ceased. All three crops were affected and in the *tal* area further damage was done by insect pests. The people, harassed by the influenza epidemic and the high prices of imported articles, with their stock of food-grains at its lowest, were thus confronted at the end of the year with a general failure of crops. In Jamui and Kharagpur thanas arrangements had to be made for the distribution of gratuitous relief and agricultural loans were distributed on a liberal scale. The able-bodied men emigrated in large numbers, in many cases leaving their families behind them to be supported by gratuitous relief. The total number of deaths in the year was 131,000 of which fever and influenza contributed 98,000. The deaths were 50,000 in excess of births and the death-rate rose to 61·2 per mille. In 1919 it fell to 40·7, but the birth-rate dropped to 30·5 from which it again rose slightly to 31·5 in 1920. The monsoon of 1919 was a specially good one and the fine crops that resulted from it went far towards re-establishing the normal economic state of affairs. For the whole ten years the recorded births were well in excess of the recorded deaths, by 39,000 in the case of males and by 26,000 in the case of females. The census showed that the loss of population in this district was 105,534 or 4·92 per cent., only 573 less than the loss in Bhagalpur which was the heaviest in the province. There was a large gain in Jamalpur thana, and a small gain in Monghyr owing to the extension of the railway workshops at Jamalpur. Otherwise there was a loss in every thana in the district. In the sadr subdivision the loss was heaviest in Lakhisarai. In 1911 a large number of labourers were engaged in cutting the crops in Lakhisarai thana at the time of the census, but in 1921 the crop-cutting had been almost completed before the census was taken, so that the loss recorded here exaggerates the loss to the permanent population : there was however some real loss, for the total number of recorded deaths in this thana during the decade exceeded the total number of births.

The loss of population was heaviest in the Begusarai subdivision (10·66 per cent.), which is attributed by the Census Commissioner to the fact that the density of population had here reached a point which was higher in proportion to natural resources than that reached in other neighbouring tracts, with the result that as other tracts developed the pressure in this area was relieved.

Migration.

The population enumerated in 1921 included 71,416 immigrants from other districts, while 223,544 natives of the district were found elsewhere. A large number of the emigrants were found in contiguous districts; but except in the case of Bhagalpur, Purnea and the Santal Parganas the excess of emigrants is not very marked, and Gaya is the only adjacent district from which Monghyr receives more than it loses. Emigration is most active from the south of the district, where the infertility of the soil in a large part of the Jamui subdivision forces the people to find employment elsewhere, particularly in the Manbhum coalfield and the tea gardens of Assam.

There is also a movement of the population to the north of the district, and every year, in the months of January and February, large numbers cross the Ganges to graze their flocks in *pargana* Pharkiya, a tract of low-lying country, which is mostly flooded in the rainy season, but affords excellent pasturage in the drier months. Migration from village to village is rare, except along the banks of the Ganges and its affluents, where the frequent shiftings of the river beds necessitate the movement of the villagers. Migration of a temporary nature from the neighbouring districts of Bhagalpur, Darbhanga and Muzaffarpur into *pargana* Pharkiya is also common after the rains, when a large number of cultivators come with their cattle and till land at a low rate (*dohta* cultivation), returning to their homes when the *rabi* crops have been cut.

Towns and villages.

There are six towns in the district, Monghyr, Jamalpur, Khagaria, Lakhisarai, Sheikhpura and Lakhisarai, which in 1921 had an aggregate population of 112,845 persons. The remainder of the population is contained in 2,768 villages.

LANGUAGE.

The language prevalent in the district is Hindi. It varies from more or less pure Hindi, with a vocabulary of Prakrit-sounding words, now only to be heard among the *purohits*, to the much altered Bihari dialects of Maithili and Magahi, which constitute the ordinary speech of the people. With few exceptions, they use the latter among themselves, though even the most rustic can usually understand the more correct Hindi spoken by the educated classes in the towns. In the Begusarai subdivision, and in the greater part of that portion of the Monghyr subdivision which lies north of the Ganges, Maithili is spoken in the form classified by Dr. Grierson as Southern Standard Maithili, a dialect which is not quite so pure as the true Standard Maithili of Darbhanga. In the south of the Gogri thana, and in the eastern portion of the Monghyr subdivision south of the Ganges, in what is known as the Kharagpur country, a variety of Maithili is spoken called the *Chhika-Chhiki* dialect. It closely resembles the dialect common in Bhagalpur, and its chief peculiarity is that a sound resembling "o" (as in the English word "hot") is added at the end of words. This sound is represented in transliteration sometimes by the letter "o" as in *hamaro*, my, and sometimes by "a," as in *chalala*, he went. In the remainder of the district, the western portion of the south Gangetic tract, the main language of the people is the Magadhi or Magahi dialect of Bihar, which is the same as that spoken in Patna and Gaya.*

Broadly speaking, however, the river Ganges marks a linguistic boundary, Maithili being prevalent to the north and Magahi to the south. There is this further distinction, moreover, that the speech of the people in the north of the district bears more traces of its Prakrit origin than that to the south, and the people are less able to understand words of Persian derivation. The southern portion of the district has been more exposed to outside influences, and the people have not succeeded in preserving such purity of speech.

Other languages are not common. Santali is spoken by the Santals in thanas Chakai, Kharagpur and Jamui. Bengali is used by the Bengalis, mainly clerks resident in the district, and Urdu is confined to the towns. The character in common use is Kaithi, which, since Hindi was adopted as the language of the law courts, has largely superseded the Persian character, though the latter is still frequently used, especially by men from up-country.

RELIGIONS. Hindus.

No less than 1,831,181 persons or 92 per cent. of the population are Hindus by religion. Here, as elsewhere, the Hinduism of the uneducated masses shows signs of an

* G. A. Grierson, Linguistic Survey of India, Vol. V., Part II, 1903.

animistic leaven, as an instance of which may be mentioned the worship of Jaydeb Dube. This is a *bhut* or evil spirit, of which the following account is given in the Bengal Census Report of 1901. "Of all male *bhuts* the most dreaded is the Barham, or Brahma Daitya, the spirit of a Brahman who has died a violent death. Such spirits are specially powerful and malicious. Sometimes they are represented as a headless trunk, with the eyes looking from the breast. They are believed to inhabit large trees by the side of a river or in some lonely place, whence they throw stones at travellers and lead them astray on dark nights; and woe betide the unfortunate who should give one of them cause for offence, by unwittingly felling the tree in which he has taken up his abode, or who was in any way responsible for his death. He can only escape the evil consequences by making the Barham his family deity and worshipping him regularly. In Bihar he often becomes the tutelary deity or *dihwar* of the whole village. The worship is usually performed under the tree, usually a banyan, which he is supposed to frequent. The trunk is painted vermilion and a mound of earth is erected, on which are placed clay figures of horses or elephants, and offerings are made of flowers, betel-nuts and the like. The worship is conducted by a special priest called the Bhakta, who is not necessarily a Brahman, and occasionally he is inspired by the spirit and utters prophecies, which are implicitly believed in by the devotees.

"Some Barhams are more famous than others, one of the best known being Jaydeb Dube, also known as Bhay Haran (fear dispeller), whose shrine is at Dadri Asthan in Monghyr. It is said that some four hundred years ago he cured a Khetauri Raja, who lived at Dadri, and was in consequence given a grant of land there. This was subsequently resumed by force, whereupon Jaydeb committed suicide, and his spirit at once began to afflict the Raja. He is now worshipped by Hindus of all classes when suffering from disease, or desirous of male offspring, usually on a Monday; offerings are made of goats, cloth, sweets, milk, fruit, etc., which are taken by the priest, a Brahman."

Other curious beliefs are reported from the Chakai thana in the extreme south. One godling, known as Garbhu Kumar, is said to have been originally a Goala, who was carried off by a tiger. No trace of the body could be found; but by chance a Naiya set fire to a heap of leaves, under which the mangled body happened to lie. As it is the son who sets fire to a funeral pyre, the Naiya was taken by Garbhu as his son, and one night was spirited away. Since then, both have been roaming over the world as evil spirits, whom the Goalas propitiate by offerings of goats and the Naiyas by offerings of fowls. There is a group of godlings called Karu Das, Lachman Das, Sibpat Das, Kalyan Das, four brothers, who became *fakirs*, and went away to some unknown place. They are supposed to visit the earth at times and are always welcome; but they cannot bear sight of women, and so, when they are worshipped, all women and girls are kept away. The offerings consist of *ganja*, *bhang*, milk, etc.—all articles liked by *fakirs*.

In the same tract Hindu females observe certain special ceremonies, such as Barshat, the worship of Siva at Jeth Amawasya, to ensure the long life of their husbands. At Mauna Amawasya women sit under a *pipal* tree on the day of a new moon, provided it is a Monday, keeping silent (*mauna*) the whole time. This is supposed to cure any constitutional infirmities which may prevent conception. Little girls also have their special worship, such as Nag Pach to propitiate the serpent god, observed on the Sukla Panchami of Sraban, when milk and fried paddy are offered and girls dance and sing; Yamdwitiya to propitiate Yama, the king of death, and so preserve the lives of brothers; and Ganesh *puja* to ensure the prosperity of future husbands.

Here too a deity called Jaksh, or simply Naiya, is regarded as the protector of the village like the *gram devata* or village godling elsewhere, men of all castes and creeds believing that the ghost of some particular Naiya protects each village from epidemics of disease and the depredations of wild animals. The Jaksh or Naiya is worshipped under a big tree just outside the village site and is represented only by an earthen mound. Both Hindus and Muhammadans propitiate him with offerings of eggs, sacrifices of goats, fowl and pigs, and libations of liquor; for it is held that if the Jaksh is not propitiated regularly, epidemics will break out and the cattle will be devoured by wild animals. The worship takes place only on one day in the year between Asarh and Bhadra.

Muhammadans. Muhammadans number 189,020 or 9·3 per cent. of the population. They appear to have been attracted to Monghyr at an early period, the town being made one of their seats of government. Smaller settlements were also scattered over the district round a number of chiefs, who received *jagirs* from the early Muhammadan emperors and kings. The Muhammadan centres are still particularly noticeable, especially Monghyr, Sheikhpura, Husainabad and Ballia Lakshminia. On the other hand, it is said that even a casual observer is struck by the fact that the faith of Islam has not spread much in the district as a whole, and that the local Muhammadans have adopted several of the customs and manners of the Hindus, the chief among them being the prohibition of widow-marriage. Except among higher castes such as Brahmans, Babhans, Rajputs and Kayasths, widow-marriage is practised among the Hindus; but the Muhammadans of Monghyr interdict women from marrying a second time, and those that transgress the custom are liable to be socially banned. The most numerous sections are Sheikhs numbering 72,324, Jolahas (32,343), and Kunjras (27,774).

Maulvi Nazir Husain, a Muhammadan reformer, who inaugurated the Ahl-i-Hadi movement in North Bihar, was a native of this district, who made his home in Delhi. The members of the Ahl-i-Hadi sect, it may be explained, are the modern representatives of the Wahabis.

Animists. At the census of 1921, 15,107 Santals were returned as Hindus, and 7,970 as Animists. There have been curious variations in the numbers returned as Animists or Santals by religion since 1881, which are probably due to the idiosyncracies of the enumerators, to the different constructions placed on the orders regarding classification, and to the difficulty of distinguishing Animism from the lowest forms of Hinduism. However this may be, it is a well-known fact that, though there is no open proselytism, the Santals are being affected by intercourse with their Hindu neighbours and have appropriated many Hindu religious customs and ideas.

The following account of the religion of the Santals is quoted from *Santalia*, by the Revd. J. M. Macphail, M.A., M.D., a missionary stationed at Bamdah in this district. "The religion of the Santals is little more than a fetish. They are demonolaters. When things are going well with him, the Santal is very little concerned with religion at all; but when any trouble comes, when there is severe illness in the family or an epidemic in the village, when the cattle are being carried off by disease or the crops are dying from want of rain, the Santal believes that these calamities are due to the ill-will of the demon-gods, whose anger must be appeased by means of sacrifice. He offers up his fowls, sheep, goats, buffaloes, and as late as 1878 a human sacrifice is known to have taken place. On the outskirts of each village is a grove of *sal* trees, often the finest trees in the neighbourhood, for they are never allowed to be cut down. This grove is sacred to the principal demons or *bongas* (Jahirera, Moreko Turuiko, Marang Buru, Gosaiera and Pargana Bonga), each of whom is represented by a piece of quartz, daubed with red paint, at the foot of one of the trees. In this grove, at festival times, the men of the village assemble—for the women have nothing to do with religion—and sacrifices are offered by the village priest in the name of the community. Combined with this there is a form of religion in some respects similar to the ancestor-worship of the Chinese. In the main street of the village a rude shed is found, a thatch supported by five posts, one in the centre and four at the corners. At the foot of the central post is a stone or roughly carved piece of wood, sacred to the *manes* of the deceased village chief or headman. Here, too, sacrifices and offerings of various kinds are offered by the villagers. But these by no means exhaust the list of demons. Everything mysterious, everything disastrous, is due to demon agency. *Bongas* people earth and air and water, and every family has its household demon as well, whose name is kept a secret by the head of the house till just before he dies. Then he whispers it to his eldest son. The object of this secrecy is not to give offence to other demons, who are all consumed with jealousy, by letting them know which demon is specially honoured with this family's patronage. When preaching to a crowd in a Santal village the writer once pointed to the stocks and stones, to which the people had recently been sacrificing, and asked them if they really thought these things could do them any good. 'No', said an old man in the crowd, 'they can do us no good, but they can do us a great deal of harm'. That, in a word, is the sum and substance of Santal theology, their body of divinity. "

A report from the Chakai thana specially mentions the worship of three deities, viz., Bhumphor, Jahirera and Khutadanta, by the Santals in the locality. According to that report, Bhumphor is a spirit who protects the people in the village itself; Jahirera protects them in the jungle; and the worship of Khutadanta is said to ensure success in hunting. Bhumphor is worshipped in a hut inside the *basti*, being represented by two stones on an earthen mound, and receives offerings of goats, fowl and liquor. Five stones placed under a *sal* tree represent Jahirera, to whom offerings of sheep, pigs, goats, fowl and liquor are given. For the worship of Khutadanta a Santal applies oil to a *sal* leaf, chants some *mantras*, and names the trees or stones haunted by spirits, this ceremony being called *chahan*. Offerings of fowl and liquor are made near all such trees and stones. Both Bhumphor and Jahirera are worshipped on four days in the year in the month of Pus, Phalgun, Magh and Asarh. Khutadanta is worshipped once a year in each jungle on a day in Baisakh or Jeth.

Christians.

Christianity has slowly advanced in Monghyr during the last forty years, the number of its adherents rising from 1,091 in 1881 to 1,433 in 1901 and to 1,649 in 1921. Of these 606 are Indians, and it is noticeable that proselytism has been most successful in Chakai, where the number advanced from 11 in 1872 to 169 in 1891 and to 293 in 1921.

Christian Missions.

There are two Christian Missions at work in the district, the Baptist Mission and a branch of the United Free Church of Scotland Mission to the Santals. The Baptist Mission was established at Monghyr in 1816, and the translation of the Hindi New Testament at present in use is the work of one of the Monghyr missionaries. There are sub-stations at Begusarai, Lakhisarai, Khagaria and Kajra. The women's branch of the mission was opened over forty years ago. It is under the supervision of a European lady-missionary, with Indian Bible-women and teachers, who manage some primary schools for girls. There is a local Baptist Christian community at Monghyr, with two church buildings, in which services are held regularly in Hindustani and English. The missionary in charge is also Minister of the " Union Church " at Jamalpur, and he conducts the Sunday evening service there.

The United Free Church of Scotland Mission has a branch at Bamdah in the Chakai thana. This Mission was established in Pachamba in Hazaribagh in 1870, and extended its work to Chakai in 1879. A hospital was built at Bamdah in 1894, which has since been much extended and is now worked by a graduate in medicine. Besides this, a branch dispensary at Chakai Bazar is worked by the Mission with the help of a grant from the District Board, which erected the building. A considerable number of schools have also been opened by the Mission, including several night schools, where boys and field labourers learn the rudiments of education after their day's work is over. With the exception of a training school for preparing boys to work as teachers, which teaches up to the Upper

Primary standard, and a similar institution for girls, the other schools are small *pathshalas* in the scattered Santal villages. The converts are, with a few exceptions, Santals.

TRIBES AND CASTES. **Aboriginal races.**

Tradition points to a former predominance of aboriginal races, which were ousted by Rajput immigrants. The northern portion of the district is believed to have been long under the sway of the Bhars, traces of whose rule are still found in ruined forts in the Begusarai subdivision and in *tappa* Saraunja to the extreme north-east. The Pharkiya *pargana* again is said to have been held by lawless tribes of Dosadhs, who were eventually subdued by a Rajput soldier. The south was comprised in the Jharkhand of the Muhammadan historians, a broad belt of forest and rolling country, which, beginning at the Son and stretching as far as Midnapore, was the boundary of and barrier against early Aryan invasions. Tradition relates that in the Kharagpur *pargana* the original inhabitants were Khetauris, who were ruled over by 52 chiefs, until they were overcome by three Rajput brothers, the founders of the Kharagpur Raj, who had taken service with one of them. To the south of Kharagpur the country was similarly seized from the Khetauris by the Rajput ancestor of the Rajas of Gidhaur.

The south of the district is still the home of castes and tribes of aboriginal descent, such as Santals, Bhuiyas, Kols and Koras, among whom the Santals alone number 23,000, almost all found in thanas Chakai, Jamui and Kharagpur. As elsewhere in India, the aboriginal races have been driven back by successive waves of conquest into the hilly tracts, where it was impracticable to follow them up, and whence they have subsequently extended themselves again towards the plains. There they have come into contact with the Hindu community, and considerable commingling has resulted. Ethnologically, the difference is still easily perceptible. The features of the inhabitants of the hills, with their low brows and flat faces, have a strong resemblance to the Mongolian type, and they contrast strikingly with the high-browed Aryan-featured Babhans of the northern parts of the district.

The Khetauris.

The Khetauris or Katauris* are believed by Mr. W. B. Oldham to have been the old Hinduized and aristocratic section of the aboriginal Maler, whose descendants still survive in

* Mr. Oldham spells the name Katauri, but in the returns of the last census it was spelt Khetauri, Khetari, Khetari or Kheturi, and never as Katauri. The total number returned was only 3,712, and many must have been classed as Rajput or Khatri. [E. A. Gait, *Bengal Census Report* of 1901, Part I, pp. 392-393.]

the Sauria Paharias of the Santal Parganas. He calls them the brethren and kinsmen of the Maler "who loved the ease and fertility of the plains and gave the race its kings and princes; and later on coalesced and intermarried with the Rajput Aryans as they approached, and took their Hinduism with the status of its soldier caste; and never daring to call themselves Kshatriyas at once, were found with the name of Kshatauri or Katauri, as the nearest permissible approach to it. ......The Katauris have long since disappeared from the west and south of the tract which they used to cover, and have receded towards their ancestral hills, driven back, I believe, by the Bhuiyan invasion from the south about three centuries ago, which established the petty kingdom of Lachmipur, the *pargana* of Passai, and most of the *ghatwalis* of Sarath Deoghar, and immediately preceded the formation of the sovereignties to Birbhum and Kharagpur by two Pathan adventurers." *

Regarding this theory, Mr. H. McPherson,† I.C.S., writes, after studying the unpublished manuscript of Buchanan-Hamilton and with special reference to the Santal Parganas:— "Dr. Buchanan's notes put an entirely different complexion on the disintegration of Katauri ascendancy. He shows that the princes of the Sauria Maler in the plains north and west of the hills, and in the Manjhwe valley within the northern hills, were of a caste called Nat or Nat Pahariya, who were closely akin to, if at all distinguishable from, the Maler. These Nats were not displaced till Akbar's advance about 1576. Katauri domination was confined to the region west of the hills, and the Bhuiyas were not invaders who helped Pathan adventurers to overthrow the Katauris, but were the aborigines of the forest tracts, over whom the Katauris exercised sway, and to whom they were closely related by blood or inter-marriage. The Katauris had possession of the more open country to the north; the Bhuiyas held the forest tracts as *ghatwals* under the Katauris. The adventurers who overthrew the Katauris of Kharagpur were not Pathans, but Rajputs. The foundation of the Pathan Kingdom of Birbhum occurred about the same time as the expulsion of the Katauris from Kharagpur, but there was no apparent connection between the two events. Both occurred about the middle of the 16th century. A grandson of the Rajput conqueror of Kharagpur gave offence to Akbar

* W. B. Oldham, *Some Historical and Ethnical Aspects of the Burdwan District* (Calcutta, 1894).

† Now Sir Hugh McPherson, K.C.I.E., C.S.I.

and was treacherously assassinated in 1601. His son was taken to Delhi and there embraced the Muhammadan faith, and Kadir Ali, who was Raja of Kharagpur in Buchanan's time, was the direct descendant of this Muhammadan convert. Younger branches of the family remained Rajput and some of them received *ghatwalis*, a circumstance which gave much offence to the Bhuiya *ghatwals*.

"The proposition that the Bhuiyas were not the sixteenth century invaders who overthrew the Katauris, but the original inhabitants of the forest tracts and guardians of the passes, who owed some sort of allegiance to the Katauris and in turn to their successors, the Rajputs, is proved not only by Dr. Buchanan's historical notes, but also by his ethnical enquiries. The Bhuiyas belong to the same Dravidian stock as the hill Maler. They have lost their Dravidian tongue and have taken on a veener of Hinduism. Their chiefs made the usual Kshatriya pretensions and calling themselves Surajbansis disclaim connection with their Bhuiya kinsmen. But the physical characteristics of all are alike Dravidian, and in Captain Browne's time (1772—8) the chiefs never thought of claiming to be other than Bhuiya. The highest chiefs of the Bhuiyas are called Tikaits, and are supposed to have received the mark of royalty. Inferior chiefs are called Thakurs, and the younger members of noble Bhuiya families are called Babus."

Buchanan, repeating the tradition current in his time, says that the Khetauris appear to have lived in brick houses, and to have been somewhat more civilized than the barbarians by whom they were expelled. The same tradition still lingers. Not only is the name of the Khetauris still widely known among the people to the south, but the tradition that at one time they were the owners of the soil, is confirmed by remains attributed to them. There are ruins of brick houses at Taranhar, one mile from Bamdah, with old bricks of peculiar size and dimensions, which are said to have been built by them; and in Raja Tarai, a hamlet of Karangarh two miles from the Kakuria inspection bungalow, there are the remains of what seems to have been a fort, which the villagers allege was once occupied by the Khetauris. A curious instance of the belief in the wealth of the Khetauris is reported. A few years ago, it is said, some Khetauris came from the Santal Parganas and stayed in Taranhar for a night, during which they dug up the ground in several places searching for buried treasure. The villagers allege that their leader was a descendant of the Khetauris, who had discovered from an old paper that there was treasure buried at a certain place in the village; and as the party decamped in the dark, the villagers all suspected that they had actually found buried treasure.

PRINCIPAL CASTES.

The marginal table shows the Hindu castes numbering over 100,000, while Brahmans (82,564), Chamars (57,045), Kandus (58,635), Rajputs (56,840), Tantis (80,589), and Telis (55,151) have each more than 50,000 representatives.

Ahir	245,527
Babhan	175,781
Dhanuk	116,153
Musahar	127,814
Dosadh	118,585
Koiri	96,614

The following is a brief account of each of the six principal castes.

Ahirs.

The Ahirs or Goalas are an agricultural caste, whose hereditary occupation is that of herdsmen. They tend flocks and herds, and though most are settled cultivators, some still roam about from place to place, seeking fresh pasturage and selling milk to the Gauras, who prepare *ghi*, which is purchased wholesale by the *mahajans*. They are most numerous in the Gogri thana, where they came originally to graze their cattle on its extensive pasture lands, but stayed to cultivate. As a class they are said to be less litigious than Babhans but no less fond of a fight, and they sometimes display an aptitude for combination which is uncommon among other sections of the population.

Babhans.

The Babhans are mostly engaged in cultivation, and the characteristic occupation of the caste is that of settled agriculturists. They comprise tenure-holders of all grades and occupancy and non-occupancy raiyats, but are to a large extent petty proprietors as well as tenants. According to their own account, though ranking as high-caste cultivators, they are not particularly sought after as tenants, because they cannot be called upon for *begari* or forced labour, and also because they cannot pay a high rate of rent; for they will not plough themselves, but employ labourers for the purpose. The truth appears to be that their bold and overbearing character, and their tendency to combine in strong and pugnacious brotherhood, render them undesirable tenants in the eyes of an exacting landlord.

Dhanuks.

The Dhanuks are a servant class found in every place where there are high-caste Hindus. They perform the menial duties of the household, along with their wives and children, and are generally employed as personal servants in the households of the upper classes. Some of the Dhanuks are also cultivators, and the females act as maid-servants.

Musahars. The Musahars are field labourers, whose wages are paid in kind, according to the immemorial custom in the villages. They live in a kind of social thraldom, sometimes selling themselves, their wives, and children to lifelong servitude for paltry sums.* With an ingrained aversion to emigration, pilfering in times of plenty, and living upon roots, rats, snails and shells, they cause considerable difficulty to Government officials in times of dearth. They live apart from the *basti*, and are not so numerous in the north-western part of the district as in the south-eastern. The bulk of the caste are field labourers and palankeen bearers, and only a few have attained to the dignity of cultivating on their own account or acquired occupancy rights. The name Musahar is believed to mean rat-catcher or rat-eater, and is an appropriate designation, for they are professional rat-catchers. Mr. Lockwood, in his work *Natural History, Sport and Travel*, writes :—" The Mushirs, or Mousers, as they may be called, are found in every village of Monghyr, and are half-starved even in times of plenty. They seldom see coin, but receive their scanty wages in coarse grain, which they flavour with rats, mice, snails and jungle roots, whilst living in hovels which an English pig would consider poor accommodation. One would imagine that such persons must find difficulty in getting wives, but the contrary is the case, for bachelors and spinsters are unknown. Directly they arrive at the age of puberty they present themselves at their landlord's house, and having signed a deed binding themselves to remain in bondage for the term of their natural lives, receive a few shillings in return, with which to entertain their friends at a marriage feast, and to set up house."

Dosadhs. The Dosadhs are another caste probably of aboriginal descent. Their characteristic occupation is to serve as watchmen or *chauikdars*, but they are also employed as village messengers (*gorait*), grooms, elephant drivers, grass and wood cutters, punkah-coolies and porters. They rank in this district among the most persistent criminals known to the police, but have a good character as carriers of goods. Some of the *chaukidars* and *goraits* hold small allotments of land rent-free

---

* As Kamiyas. But see p. 133.

in return for the services rendered by them to the village; but generally speaking, their thriftless habits hinder them from rising above the grade of occupancy raiyats, and a very large proportion of them are merely tenants-at-will or landless day labourers.

Chakai Dosadhs

To the south in the Chakai thana there is a body of Dosadhs know as notorious criminals. The following note regarding this class has been contributed by the Rev. J. M. Macphail:—

"The Dosadhs of Chakai have for long been notorious as daring and adventuresome robbers. By an ancient arrangement *ghatwals* are appointed by the Maharaja to guard the pass from Bamdah to Batia, and they are supposed to accompany travellers through the pass. I believe the object of this arrangement was to protect travellers from thieves as well as from wild animals. One reason why these thieves find Chakai so convenient a centre is because it is near the borders of three other districts, Hazaribagh, the Santal Parganas and Bhagalpur. The thieves can easily do a night's work in villages (some of them important centres like Baidyanath) in these other districts, and be back to their own villages in the morning. They are chiefly found in a large village, Chandermandi, near Nawada (often called Chor Nawada), and Karangarh about five miles from Chakai thana. There is very little secrecy about their way of life, and otherwise honest and respectable people in the neighbourhood buy all sorts of articles from them, knowing quite well that they have been stolen. They are not very particular regarding the articles they steal, although perhaps they may be said to have a partiality for brass vessels. Foodstuffs, clothing, and standing crops are frequently stolen; and they are also cattle raiders, sheep and goats at any rate being very frequently stolen.

"They are remarkably expert. I have known of cases in our own Mission Hospital here, where they have stolen the bedding on which a man was lying, and also the gold spectacles he was wearing, without his being conscious of the fact at the time. I have been told that they will steal a *lota* from the midst of a group of travellers camped round a fire at night. Their *modus operandi* is to approach the group, asking to be allowed to get a light from the fire, and to take the opportunity of letting a noose fall over the neck of a *lota*. The *lota* is then jerked away, when nobody is watching it. They are also very daring. I have known them steal the clothes from off the body of a pilgrim woman by sheer force, leaving her practically naked, in the verandah of our hospital, on a dark night. An illustration both of the daring and the expertness of the thieves was afforded some time ago when they dug through the walls of the Chakai thana and killed the Sub-Inspector's horse.

"Hitherto the Dosadhs have also shewn remarkable cleverness in evading the law. Sometime ago a Santal in Bamdah found a Dosadh, one of the leaders of the gang, stealing a sheep from his court-yard at night. He raised a hue and cry, and gave chase, and the man was caught and taken to the thana. In due course the trial came on at Jamui, but the thief was acquitted and the man who caught him was prosecuted for assault or on some similar charge. Another fact that has helped to secure their immunity has been that there has been too little hostility to them and their doings on the part of their neighbours, who do not object to their stealing other people's property, and who also derive some profit by buying goods from the thieves at less than cost price. It used to be the custom here, as in many other places, to employ Dosadhs as village *chaukidars*, and the people who remember that time tell me that when anything was stolen, if they made sufficient noise about it, they always got it back through the *chaukidars*."

Koiris.

The Koiris are agriculturists pure and simple, but they are distinguished from the Kurmis and other purely cultivating castes by their skill in rearing tobacco, and other special produce requiring more careful cultivation than the staple crops. In the neighbourhood of large towns they work as market-gardeners. Many Koiris are tenure-holders, and here and there members of the caste have risen to be petty zamindars; but most of them are prosperous cultivators, holding occupancy rights.

ECONOMIC CONDITIONS.

Writing in 1875, the Collector reported that the "main body of the people are not much to be envied, nor can they be said to be prosperous, since they get barely sufficient food and clothing. At the same time, so very little suffices to keep the villager well and strong, that any person in the possession of five rupees per month, and blessed with good health, may be said to be prosperous and happy, even though he has a family to support." This account still holds good to a certain extent, though the standard of living has risen, as evidenced by the desire for better food, better clothing, and generally a better mode of living. Gold and silver ornaments are more common; brass utensils have largely usurped the place of earthen pots;

shoes, umbrellas and finer clothes are more extensively used; and articles of food which were formerly considered luxuries are now commonly consumed. Moreover, though wages are low, the wants of the labourers are few and many of them are aboriginals or semi-aboriginals, who require little to make them happy. A cultivator's hut can be run up for little more than ten rupees; while if he is a popular man, and can get his neighbours to give him a helping hand, it will cost less. His clothes cost him a mere nominal sum; shoes and stockings he does not require. His wants in the way of furniture are equally modest, and he has no use for a chair, finding the ground a more comfortable seat. Above all, he can get a good meal of rice, fish and spices, with a cucumber or melon to finish up with, at a trifling cost.

Generally, the people live abstemiously and with prudence and forethought. But here, as elsewhere, when a religious or social ceremony has to be performed, they not only recklessly squander their hard-earned gains, but often involve themselves in debts which they never succeed in throwing off. As an ordinary instance of this may be mentioned the case of a well-to-do cultivator who died leaving about Rs. 600 in cash and grain; his son squandered the whole of this sum in entertainment at his *sradha*.

Houses.

The ordinary dwelling house of a poor cultivator costs very little, for it generally consists merely of mud walls with a thatched roof. Many have the bamboos required on their own land, and they use as thatch their paddy straw or thatching grass, or purchase them at a low price from their neighbours. Inside there is very little except a few baked earthen vessels, a wooden box, a brass plate or two, and a hand-mill for grinding rice. The small shopkeeper's house similarly has mud walls, but sometimes a tiled roof. The furniture consists merely of a few large wooden chests with rude locks, in which the commodities he deals in are stored; two or three smaller chests for his own property, in which he keeps any valuables or good clothes that he may have; a couple of cane or bamboo stools, 6 or 8 inches high; rough beds, of netting on a wooden framework; a hand grain mill; a few brass pots and pans; and usually a spinning wheel for his wife.

Dress.

The dress of the peasants is simple, that of the men being usually confined to a waist cloth (*dhoti*) and a head cloth (*pagri*), which is used in the hot weather to keep off the heat of the sun and in the cold weather to protect them against cold.

They sometimes also wear a cloth over their shoulders, when they wish to appear more respectable than usual. These articles cost but little. Umbrellas are now largely coming into use, and coarse woollen shawls are also worn during the winter. A *sari* and *kurta* (bodice) are the dress ordinarily worn by females of the lower classes.

Food.

The cultivators generally live on paddy and yams for three months, Pus December) to Phagun (February), on *rabi* for six months, Chait (March) to Bhado (August), and on *bhadoi* for the remaining three months, i.e., Asin (September) to Aghan (November). The food of the people in general is of the coarsest and cheapest kind procurable. Paddy is eaten in two ways, either boiled as rice or husked as *chura*, the latter being the most common form. Yams, either boiled or fried, are very largely eaten by the poorer cultivators and labouring classes on account of their cheapness. The destitute generally obtain some yams by digging in fields from which the crop has been removed, as much as five seers a day being gleaned at times. Then comes the *rabi* season. Barley is the principal *rabi* crop that is consumed by the people on account of its comparative cheapness. It is mostly eaten after being ground into *sattu*. The *sattu* is either made into gruel and seasoned with chillies and salt, or baked into *chapattis*. Wheat is eaten by the better class of cultivators. Gram, *rahar* and *kerao* are also eaten by the people either ground, boiled, or fried. Then comes the *bhadoi* season. Of the *bhadoi* crops *marua* and maize are largely eaten by the people and are extensively grown. *Marua* is eaten only in the shape of *chapattis*, but maize is consumed in various ways, parched, boiled, or fried. Of vegetable crops, pumpkins and brinjals are largely eaten. Potatoes and *parwals* are consumed only by the more prosperous. Chillies are grown and used in large quantities, and *dahi* (curds) is a luxury eaten with relish. It is noticeable that, although more rice is grown than any other species of grain, the mass of the people eat mainly preparations of wheat, barley, Indian corn, and other grains. The upper classes even do not eat nearly so much rice as the Bengalis.

The poorer classes have generally two meals a day; breakfast at noon, consisting of bread or parched grain, and dinner at seven o'clock in the evening. Besides food-grains, many varieties of vegetables are eaten by all classes, and several wild plants are gathered as *sag* or greens. The flowers of the *mahua* (*bassia latifolia*) are an important item in the daily meals of the poorer classes in the southern portion of the district, men and animals being alike fond of them. The flowers, or strictly speaking, the succulent petals, fall in profusion in April and May and are then collected and dried. In the north of the district mangoes are in equal request, while the jack, guava, and plantain play a subordinate part.

Fish are also eaten largely, the Ganges and the numerous marshes in the north-east of *pargana* Pharkiya supplying vast quantities. The siluroids, being generally reputed unclean feeders, are not in much favour with any except the lower classes. Carp, *hilsa* and *moh* are chiefly eaten by the upper classes, and the mullet is considered a great delicacy. As it is difficult to take this last fish with the net or rod, it is mostly speared in the rivers or shot in the marshes, as it springs into deep water out of the shallows among the reeds. The rivers form a vast fish preserve; there appears little fear of the supply falling short, although no close season is known, and the meshes of the nets used are without restriction as to size. River turtles are eaten by the lower classes, as are also land and water crabs, which are very common. The burrowing land crab affects the sides of the marshes; and in *pargana* Pharkiya, during the cold season, its little mud forts are met with everywhere. The Musalmans eat beef, and goats' flesh and mutton are in favour with most classes. The Musahars catch and eat the field rat, which, feeding as it does on rice, is not unpalatable. The Santals and other forest tribes are said to eat almost any living thing they find in the woods. Birds, with the exception of the domestic fowls, do not enter largely into the food supply, as they are more difficult to procure than mammals and fish, but the lower classes will eat almost any bird they can catch.

Famine foods.

The following list of the roots eaten by the lower classes of the people, which form a valuable addition to their food supply in time of famine, was drawn up during the progress of the famine of 1874 by Major Waller, then District Superintendent of Police. (1) *Putal koura* (*batatas paniculata*), a creeper which grows largely in the valleys between the hills and on the banks of streams. Its roots grow to a very large size and are much used throughout the south of the district. When small, they are eaten raw, but the larger ones are boiled. They are insipid in taste, but wholesome. (2) *Gaithi* is a species of *dioscorea* growing near the hills; its root is cut in pieces, boiled till quite soft, and then eaten; it is sweet to the taste. (3) *Tamuli* (*curculigo orchioides*), a small plant with a leaf like the turmeric, having a single root, which is eaten boiled, and has a rather pungent flavour. (4) *Sutawar* (*asparagus sarmentosus*), the root of a small prickly creeper found among the hills, which is boiled and eaten; its taste is insipid. (5) *Khela kheli* is a root obtained from the Kharagpur jungles; it is eaten boiled, and has a rather astringent taste. (6) *Siyah Munsli* (*mudrannia scapiflora*), a small plant with a leaf like the ginger and a single root, which is eaten boiled. (7) *Kand*, probably a species of arum, has a root resembling the *sakarkand*, and of the same taste; it is found in the Kharagpur hills. (8) *Asar*, a creeper found in the Jamui hills, with two or three tubers under each plant, which are boiled and eaten. (9) *Moronara* or *algogi*, a small creeper met with in the hill tracts, having three or four tubers under each plant, which are boiled and eaten, and have an insipid taste. (10) *Bongo* is also a small climbing plant found in the hills, with a single root, which is boiled and eaten. (11) *Ijwar*, a small tree common in the south. The bark is taken off the root, which is scraped, and the inner part boiled; its taste is rather earthy. (12) *Piska* is the root of a creeper, also found in the southern hills; it is boiled all day, then steeped in cold water all night, and again boiled next day, when it is fit for use. (13) *Kolo*, a species of *dioscorea*, which is very common in the southern jungles, four or six pounds weight of tubers being found under each plant. Several hundredweights are dug up and boiled together, as the process of cooking is troublesome and much the same for a large or small quantity. The Santals who use this root to a considerable extent, say that the steam which issues from it when being boiled is very intoxicating; and the person attending to the boiling, when moving or touching the vessels, either covers his mouth or turns away his head. The tubers are considered poisonous, if eaten raw or before being well cooked. The roots are sliced and boiled all day, then steeped in cold water all night, being generally put in baskets and sunk in a running stream; the next day they are reboiled and again washed several times, after which they are fit for eating. The favourite way of serving them is as follows. After they have been boiled and washed, they are mixed with *mahua* or other fruit, and made into balls, which are ready for immediate use, or may be warmed up again. (14) *At* or *biran* is the root of a creeper found at the foot of the hills, weighing from four to six pounds. This plant also bears a small round fruit, which is used for food. The root is peeled and boiled till quite soft, and has a pleasant taste. (15) *Kakori* (*momordica dioica*) is a creeper common in Chakai; there is a single root under each plant, which is boiled, and then peeled before eating; it is sweet and pleasant to the taste. (16) *Tona* is another creeper found in the hilly tracts towards the south; each plant has a single root, which is boiled whole before being eaten; it is pleasant to the taste, and is said to be a wholesome, strengthening food. (17) *Gorkhundi* is a small tree found in the hills, which bears a small sweet fruit; the root is cooked by boiling, after which the thick bark is stripped off and the inner part eaten. (18) *Musla simul* is the root of the young *simul* or cotton tree, and is very commonly used as food by the jungle people in the rainy season, when it is tender and juicy. It is also regarded as a strengthening tonic; for medicinal purposes it is dried, ground into powder, and mixed with milk and sugar. The root, when used as an article of food, is boiled and the thick bark removed. (19) *Boruni* is a small plant found throughout the hills, and bears a round fruit which is used as a vegetable; it is eaten boiled. (20) *Arwa* is the root of a creeper found in the same localities, which is also eaten boiled. (21) *Kand bisara* is the root of an aquatic plant found in the Kabar Tal and other large marshes north of the Ganges; it is boiled, peeled, and then eaten. (22) *Karhal*, probably a species of *Nymphœa*, is common in the north of this district and is very generally used. As the water of the marshes recedes and leaves the land dry, these roots, which are of considerable size, are dug up, boiled and eaten; they are pleasant to the taste. (23) *Kana gijari* and *lauk* or *loka* are similar roots, common in pargana *Pharkiya*. (24) *Surki* (*nymphœ lotus*) is the root of a very common water-plant with a red flower; it is obtained in a similar manner. It is eaten boiled; its taste is slightly pungent. The seeds of the flower are also collected and made into a sweetmeat called *koi-ka laddu*, found in most confectioners or *halwais*' stalls. (25) *Kauchur* (*cyperus*) is the root of a small sedge. (26) The root of the *bauhinia vahlii* is also eaten, but has no specific vernacular name, being merely called a *sag* or vegetable. (27) *Chichor*, the local name for a marshy sedge (*cyperus palustris*), the roots of which are dug up in the dry weather. A man may collect about six pounds a day, with average luck and labour. They are usually ground into a kind of flour, and made into bread or cakes, but are also eaten raw. These sedges have been described as "famine thermometers," for in times of plenty they are abandoned to the pigs, but in times of scarcity the bulbs are grubbed up for food; and, writes Mr. Lockwood—"When I saw 500 persons *chichor*-hunting on a single marsh, I knew that there was famine in the land."

VILLAGE OFFICIALS.

The following is a brief account of the village officials found in the district, from which it will be apparent that many of them are merely zamindars' servants and that the old communal life of the village has almost disappeared.

Patwari.

The *patwari* was formerly one of the principal officers of the *kanungo's* staff, and, as such, a subordinate official of Government. His chief duty was to check the proceedings of the farmers of the revenue, at the same time, he was the repository of information concerning the village lands, their crops, and boundaries. He now keeps the rent-roll of the village, and is merely the village accountant. He has long ceased to perform any public duties as an officer of Government and has become nothing but the servant of the zamindar whose accounts he keeps. The *jama-wasil-baki* (demand, collections and balance) accounts of every raiyat in the village, with the quantity of land enjoyed by each, and other particulars, are kept by the *patwari*. The appointment, suspension and dismissal of the *patwaris* rest practically with the zamindars, though there are occasional appeals under Regulation XII of 1817; but that regulation is practically a dead letter in Monghyr, the appointments made by the zamindars being but rarely reported. The office is hereditary, provided a worthy member of the family is forthcoming, and provided the incumbent is not dismissed with disgrace.

The *patwaris* of Monghyr do not enjoy a higher reputation for probity than elsewhere, and are, as a rule, ignorant and unprincipled; but they are men of power in the villages. and, being very shrewd and having much influence for good or evil, are regarded as authorities on village affairs. They generally belong to the Kayasth or writer caste, and draw pay varying from two to fifteen rupees a month, besides a small quantity of grain from each raiyat at harvest time, this gratuity being called *nocha*. They are also supposed to get an allowance from each raiyat, known as *neg* or *hujitana*. Ordinarily there is a *patwari* to each estate, but where the estate is large or divided among several sharers, there may be more than one; on the other hand, where there are several small adjoining estates under one proprietor, one *patwari* is considered sufficient for all. He sometimes has an assistant known as a *naib patwari* or *mutasaddi*.

Jeth-raiyat and *mukaddam*.

The *jeth-raiyat*, or, as he is called in the country north of the Ganges, the *mukaddam*, is the village headman, who is sometimes also designated *mandal* or *mahto*. He is the agent through whom rents are collected, and is paid a small percentage on the amount actually collected. No security is taken from him, but, as a rule, he is the most substantial raiyat in the village. The post is not hereditary, but generally descends from father to son, if the latter can maintain his position. His general duties are to carry out the orders of the zamindar, to give assistance in making settlements of land, to look after the embankments and irrigation works (if any), and to protect the village boundaries. He also arbitrates in case of differences between the villagers. His power over the other raiyats is recognized by the landlord, and he obtains his land at a more favourable rental than others. He is well versed in everything that concerns the village, and is supposed to know the value of the fields and their boundaries.

Gumashta.

The *gumashta* is merely a servant of the landlord, who has to collect the village rents, the *patwari* keeping the accounts for him. He receives a small salary from the landlord and presents from the villagers like the *patwari*.

Amin and *salis*.

A regular *amin* is employed only by the larger landlords as a permanent servant. In those parts of the district where produce-rents prevail, he is frequently a respectable villager, who measures and surveys the village crops. The man who assists him by plying the measuring rod or rope is called a *jaribkash*. As soon as the *amin* declares the area of any field, another man, called a *salis*, determines its produce per *bigha*, and assigns the share payable to the landlord. The latter name means arbitrator, as the *salis* is supposed to arbitrate between the zamindar and raiyat regarding the produce of the fields; he is remunerated by a small percentage of grain from each party. Functions similar to those of the *amin* are discharged by the *mirdaha*, who also measures and surveys village lands at harvesting time in connection with the *danabandi* system.

Badhwar.

The *badhwar's* duties are to watch crops and prevent cattle straying on them. He is paid a pittance by the raiyats in grain at harvest time. The term is derived from *badh*, i.e., the area under field crops.

Barahil and *gorait.*

The *barahil* and *gorait* do the work of peons or messengers, bring the tenants to the managing office of the estate when they are wanted, look after the crops and the use of irrigation reservoirs, and assist the *jeth-raiyat* or *patwari* in collecting rents. They are servants of the estate, and are mostly paid by *jagirs* or receive a salary in cash of one or two rupees a month.

SOCIAL INSTITUTIONS.

The social institutions established in the district are for the most part of a sectarian nature, having for their object the amelioration of the condition of the section of the community which they represent in support of the needy, widows and orphans. The principal associations of this classs are the Kayastha Sabha, the Anjuman Himayati Islam, the Bhumihar Babhan Sabha, the Gorakshini Sabha, the Whitty Relief Association and the Dixon Charitable Society. The first of these pays the school-fees of some orphan boys and maintains a few widows, and takes an interest in matters affecting the social welfare of the Kayasthas, but it does not appear to have shown much activity of late years. The second aims at the social improvement and progress of Muhammadans. It maintains a madrasa, with an orphanage attached, at Monghyr. The Bhumihar Babhan Sabha, now a practically defunct institution, maintains its identity in the form of a hostel for Bhumihar Babhan students. Another social institution is the Gorakshini Sabha, which shelters and treats a considerable number of infirm cattle; this is done chiefly under the auspices of the Marwari community. The Whitty Relief Association was started in 1924 primarily to afford relief to the flood-stricken people of the district. The Dixon Charitable Society was established recently to help poor students in the prosecution of their studies, and the needy and distressed irrespective of caste and creed. The society is maintained mainly out of donations from local zamindars.

Other institutions worth noticing are the Santan Dharma Sabha, the Orphanage and the Whitty Home for Juveniles. The first which is an organization of Hindus, founded about 40 years ago, maintains a Sanskrit Pathsala and holds religious meetings. The second is maintained by the local Arya Samaj out of public subscriptions. The third, established in

1923 with the object of giving lodging and boarding to the prisoners discharged from the Monghyr Juvenile Jail and of finding work for them, is controlled by the Juvenile Prisoners' Aid Society of Monghyr. It is maintained by public donations.

CHAPTER IV.

PUBLIC HEALTH.

THE climate of Monghyr is on the whole healthy, but differs somewhat in the tracts north and south of the Ganges. The northern portion is a low-lying alluvial tract, damp and often waterlogged in parts, while the country south of the river is higher, hilly in many places, and altogether very much drier. March, April, and May are, as a rule the three healthiest months in the year, but cholera sometimes makes its appearance, and by its ravages may cause a greater mortality than at any other season of the year. The appearance of the rains ushers in the unhealthy season. Intermittent fever, dysentery and diarrhoea, and inflammation of the lungs become more prevalent, and the two latter diseases less amenable to treatment. The changes of temperature are both great and sudden ; and those who have suffered much from malaria in previous years are apt to have returns of the disease. In November and December the weather is cool and pleasant ; and those who have been suffering from intermittent fever and dysentery gradually recover strength, if protected against undue exposure. In February rheumatism and dysentery are apt to occur among the poorly clad and whooping cough and measles may break out in epidemic form. CLIMATE.

The following table shows the number of births and deaths registered in Monghyr district during the five years 1919 to 1923 :— VITAL. STATISTICS.

——	Births.		Deaths.	
	Number registered.	Birth-rate per 1,000.	Number registered.	Death-rate per 1,000.
1919	65,035	30·5	86,781	40·7
1920	67,282	31·5	60,070	28·1
1921	82,696	4·07	63,472	31·2
1922	77,857	38·3	47,035	23·1
1923	82,483	40·6	49,280	24·2

In 1918, when the influenza epidemic was raging, the number of deaths was 130,533, and the high death-rate of 1919 is to be chiefly attributed to this disease, though the epidemic was then declining.

PRINCIPAL DISEASES.

The principal diseases of the district are malarial fevers, and kala-azar; dysentery, diarrhœa, and other forms of bowel diseases; ophthalmia, bronchitis, pneumonia, asthma, calculus, leprosy, elephantiasis, goitre, small-pox, plague and tuberculosis.

Malaria and kala-azar are very common all over the district in the rains and at the beginning of the cold weather, and a large proportion of the people bear signs of chronic malarial poisoning in the form of enlarged spleen, anaemia, and pigmentary patches on the tongue and face. Deaths from this disease are most numerous in the month of November. The returns show that fever accounted for 46 per cent. of the deaths in 1918, and for 33 per cent. in 1919; but these figures include most of the deaths from influenza in those year: and in every year many disorders that show febrile symptoms in their course, and end fatally, are grouped under this head by the chaukidars who are ultimately largely responsible for the returns. The following account of the types of fever observed in the district was contributed by Captain E. Owen Thurston, I.M.S., sometime Civil Surgeon of Monghyr:—

"Most of the types of fever met with in the district are malarial in origin. Of other varieties there is one, of which the chief feature is enlargement of the spleen and anaemia, while a low type of fever accompanied by enteritis is common in Jamalpur. As regards malarial fevers, the form most commonly met with is due to infection with the tertian parasite; next comes the quotidian variety, in which, however, the daily paroxysm of fever is often due to a double infection with this variety of parasite. A quartan infection, as far as can be ascertained, has not been observed. A typical case of fever can be shortly described as follows. The patient first of all feels out of sorts, with a little headache or anorexia, and perhaps a feeling of chilliness. The real onset begins with the ague shivering fit, which may be accompanied by vomiting, then follows the hot stage, and lastly the sweating stage and defervescence. In fact, the attack often follows the classical description of the disease. Frequently, however, the fever is remittent with a nocturnal rise and lasts a few days. This variety has chiefly been observed in the jail, where the attack has been rendered atypical by the prophylactic administration of quinine. Blood examination has generally shown the benign tertian parasite, but in other cases the malignant variety has been detected. In these latter infections, cases illustrating the comatose type have been observed. Out of a total number of 243 blood examinations in cases of fever, *Eosinphilia* was present in 18 cases, and in six cases *Ascaris lumbricoides* was found. The benign tertian parasite was present in 33 cases and the malignant tertian in 19 cases, of which seven showed crescents. The majority of these patients had taken quinine before their blood was examined. The disease is present the whole year round, but the greatest number of cases occur in October and November.

" As regards the distribution of the disease in the district—judging from the number of cases treated in the respective dispensaries and allowing for their importance and closeness to one another, the most malarious areas, in order are Kharagpur, Chapraon, Monghyr, Chaklahabad, Teghra, Begusarai, Sangrampur, Gogri, Sheikhpura, Lakhisarai and Jamui. About Kharagpur there is a range of hills with a good deal of forest, and much rice cultivation with irrigation; and southwards towards Sangrampur the same conditions obtain, except that the hills are less marked. In Chapraon there are many *jhils* or swamps, while the Begusarai subdivision is, in parts, liable to floods, but the amount of rice cultivation is comparatively small. The amount of water in Kharagpur and Chapraon is the chief factor in the prevalence of the disease, conditions being favourable to mosquito life.

"In fevers of the type characterized by anaemia and enlarged spleen, the patient is weak and emaciated, with a protuberant abdomen due to the enlargement of the spleen. His face has a peculiar earthy colour; his conjunctivæ are in extreme cases almost white or of a pale yellow colour; and he is subject to recurring attacks of fever. This type is attended by a considerable mortality, a terminal dysentery being often the cause of death. Among other complications Cancrum oris is common, and also sloughing ulcers of the leg. Whether all these cases are really of malarial origin, I am unable to state definitely, as no spleen punctures have been performed, but they approximate very closely in clinical features to cases of Leishmann-Donovan infection, and this body was found in one case in the spleen of a man dying from pneumonia, so that the probabilities are that they are cases of this disease. Another point is that they do not react to quinine, and that the proportion of these cases, compared with that of malarial fevers, is lowest where the malarial index is highest and vice versa. Another form of fever lasting up to a fortnight is characterized by enterits, and here the fever often starts moderately high and declines slowly. Blood examination has been negative in this variety, and so is the Widal reaction. Up to the present this variety has only been observed in Jamalpur and has been attributed to a defective water-supply."

Plague.

Plague first broke out in the district in January 1900, but subsided in May, only to reappear with renewed virulence in the ensuing cold weather. The total number of deaths reported in 1900 was 2,052, but, as in other districts, and as usual during the first seasons of the epidemic, the disease was far more prevalent than the reports would indicate, and much of the mortality was concealed. The parts of the district which suffered most were Monghyr town and Sheikhpura thana. Next year the epidemic was more widespread, and 4,742 deaths were returned. In 1905 plague caused 11,080 deaths, or 535 per mille. Since then plague has been an annual visitation, and as late as 1917 the death-rate from this disease was 2·4 per mille, but in more recent years the death-rate from plague has been 1 per mille or less. The disease now occurs only in the area north of the Ganges, particularly in that adjoining Darbhanga district.

Cholera.

Cholera is endemic in the district and often breaks out in epidemic form. In the epidemic of 1918 the mortality due to cholera was as high as 7·7 per mille; in 1917 it was 3·1, in 1921 2·4, and in 1919 1·8 per mille. In 1920, 1922 and 1923 the death-rate from this disease was less than one-half per mille; and it has been rare in Monghyr town since the filtered water-supply was installed.

Small-pox.

Fifty years ago it was stated in the Statistical Account of Bengal that—"Small-pox, though it is ordinarily regarded as an epidemic, is in this district, as in every other where inoculation largely prevails, in reality an endemic from which the people are never free." These remarks no longer hold good, for inoculation has ceased, vaccination has made great progress, and since the present system of vital statistics was introduced the annual death-rate has never been even 1 per mille.

Diarrhoea and dysentery. Dysentery is found at all seasons of the year, but is met with most frequently during the rains. The poorer classes suffer most from it, a fact which may be attributed to their greater exposure to the vicissitudes of temperature, and also no doubt to bad food, scanty clothing, and other privations. Diarrhoea is also met with all seasons, but is most common at the beginning and end of the rains.

Intestinal parasites. The Civil Surgeon of Monghyr, Major A. N. Palit, I.M.S., has recently made an investigation concerning the prevalence of parasites in the bowels of prisoners in the Monghyr Juvenile Jail, which shows the following result :—*uncinaria duodenalis* 81 per cent., tape-worms 75 per cent. This however cannot be taken as a true indication of the incidence of intestinal parasites in the district, as juveniles from all over the province are admitted to this jail.

Eye diseases. Conjunctivitis is common during the months of April and May, when the hot west winds are loaded with dust. It often occurs only in a mild form ; but among the poor it is apt to take the form of purulent opthalmia, resulting in total destruction of the eye or in the formation of permanent opacity of the cornea. Most of the poor show signs of trachoma. In the three years 1922—4, 5,735 successful operations for cataract were performed, mostly at the Monghyr sadr hospital or at the Mission hospital at Bamdah.

Goitre. Goitre is confined in a very curious manner to the villages on the bank of the Burh Gandak in the north-west of the district. Even animals are said to be affected by the disease in some villages.

Other diseases. Bronchitis is common in the cold weather, being chiefly met with in the old and poor. Pneumonia is fairly common. Asthma is prevalent, chiefly among the old, but it is by no means limited to them. Calculus or stone in the bladder occurs in all parts of the district, and is often found in children. Lathyriasis is fairly common in north Monghyr. Hydrocele is a very common cause of partial disability.

Infirmities. The most prevalent infirmity is blindness, 65 per 100,000 males and 59 per 100,000 females having been returned as blind at the census of 1921. It appears to be most common among castes engaged in agricultural pursuits, probably owing to the glare and dust from the sandy soil. Of deaf-mutes there are 36 per 100,000 males and 24 per 100,000 females, more commonly found along the course of the Burh Gandak than in other parts of the district, and particularly in the Teghra and Begusarai thanas.

Leprosy is not uncommon, 23 per 100,000 males and 3.6 per 100,000 females having been returned as lepers at the census of 1921. Popular belief connects the disease with general uncleanness of living and unwholesome diet, and the lowest castes most frequently suffer from it, especially the Musahars, whose unclean habits of living are notorious. The virulence of the disease also appears to increase with the descent in the social order, for the lowest classes are generally attacked with the worst form and the development of the disease is more rapid in their case. A former Civil Surgeon states that the earliest symptoms are anaesthetic patches, thickening of the ulna nerve, and slight hypertrophy of the integument of the ears and forehead. Insanity is rare, only 8.4 males and 3 females per 100,000 having been returned as insane in 1921.

VACCINATION. Vaccination is compulsory only in municipal areas but it is not regarded with any great disfavour by the people generally, with the exception of Babhans, who in this district are usually opposed to it. In 1924-5 61,261 persons were successfully vaccinated, and the average number in the previous five years was 60,584, representing 30 per mille of the population.

SANITATION. The following account of the sanitary state of Monghyr was given by the Civil Surgeon in 1868: "The main streets of Monghyr are kept clean, but otherwise the state of the town is much the same as it must ever have been. As soon as you leave the main streets, you come across heaps of putrefying filth and large holes full of fœtid fluids. Drainage is imperfect, partly because it is incomplete, partly because the levels of the drains are not correct. The rain water finds for itself a path in the centre of the roads. At intervals, when the rain is very heavy, it finds its way into a drain, which may thus be cleared out.

There are in several parts of the town hollows of great extent and of irregular shape and depth. These have not been constructed originally as tanks, but have arisen from the earth having been gradually removed for building the huts in the immediate neighbourhood, or for brick making. The description of one of these will suffice for all. It lies in an open space to the east of the town, completely surrounded by houses. It is of irregular shape, about 100 yards in length and breadth, and is partially filled with black, muddy water, the accumulation of all the surface drainage in the neighbourhood. Numbers of people are bathing and washing clothes, while the portion of the hollow not now under water is being used as a public necessary. Later in the season the water dries up, and only a fœtid black ooze is left. In the hot weather it is perfectly dry."

During the 57 years which have passed since the above remarks were written, a considerable improvement has been effected both in the drainage of the town and in the conservancy arrangements. In recent years a piped water-supply from the Ganges has been provided at a cost of over 5½ lakhs of rupees; and over two lakhs of rupees have been spent on a drainage scheme. Sanitary progress has however been more marked in Jamalpur, which is a particularly clean and well-kept town.

The following table shows the public dispensaries, including the aided dispensary at Simultala, with the number of patients treated in 1923:—

Dispensary.	Patients treated in 1923.		
	In-patients.		Out-patients.
	Number of beds.	Patients treated.	
1	2	3	4
Monghyr	73	1,064	15,033
Jamalpur	...	...	1,874
Jamui	20	272	9,942
Begusarai	18	335	14,606

Dispensary.	Patients treated in 1923.		
	In-patients.		Out-patients.
	Number of beds.	Patients treated.	
1	2	3	4
Sheikhpura	6	51	9,013
Teghra	4	52	13,593
Surajgarha	...	...	5,170
Lakhisarai	7	71	7,265
Gogri	4	52	13,978
Sangrampur	10	206	12,516
Khagaria	12	127	12,250
Mansurchack	...	...	18,823
Sikandra	6	97	6,452
Bakhtiarpur	..	...	11,813
Barbigha	...	...	10,190
Manjhaul	...	...	7,466
Simultala	...	...	3,654

The dispensary at Jamalpur is maintained by the municipality; and that at Monghyr is maintained by the municipality, with the help of a contribution of Rs. 11,500 from the district board and of Rs. 5,500 from the local Government. The district board contributes only Rs. 300 to the income of the dispensary at Simultala, but the other dispensaries named above are maintained by the Board. The most important is the dispensary at Monghyr, which was provided in 1883 at a cost of forty thousand rupees by the late Mr. H. Dear, when the old building was found inadequate for local needs. It now contains beds for 73 patients, and there are two rooms on the upper story in which European in-patients are received. It is maintained by the municipality, with the help of a big contribution from the district board, private subscriptions, and the interest on an endowment fund of Rs. 71,500. Raja

Kamleshwari Prashad Sinha made a gift of Rs. 10,000 for the construction of a ward for two paying pardah-nashin patients; and Raja Deokinandan Prashad Sinha has given Rs. 10,000 for an infectious ward of eight beds, (the Wheeler Ward, named after His Excellency the Governor of Bihar and Orissa).

In addition to these, the Bahadurpur estate maintains a dispensary, which is under the supervision of the Civil Surgeon; and there are three other private dispensaries, the Diamond Jubilee Dispensary at Gidhaur, maintained by the Maharaja, the Kharagpur dispensary maintained by the Maharaja of Darbhanga, and the dispensary at Bamdah which is maintained by the United Free Church of Scotland Mission. The medical missionary at Bamdah also visits Chakai twice a week and dispenses medicines there.* The East Indian Railway Company maintains hospitals at Jamalpur and Jhajha and a dispensary at Kiul.

The European system of medicine and surgery has steadily gained popularity, but a large number of the people have a peculiar aversion to quinine and other English medicines, which, however, may be due merely to the comparative costliness of engaging a doctor who will charge a fee for every visit. Most of them consult Indian doctors, either the Muhammadan *hakims* or Vaidyas who practise the Hindu system of medicine. In many cases, the latter two classes do not learn their craft from the study of any scientific treatises, and their therapeutics consist of a number of nostrums handed down from generation to generation. One of the principal means adopted by them for recovery consists of a total abstinence from all food or drink during the first week of an attack of fever, a system which is popularly known as *faka*. Both Muhammadan *hakims* and Hindu *kabirajes* administer drugs indigenous to the country, but some are beginning to use European medicines. Surgery is almost unknown to them, but some operations are performed by indigenous oculists and Muhammadan barbers. The former operate for cataract by depressing the opaque lens with a needle. Many of their unfortunate patients afterwards seek aid at the hospital for general inflammation of the eye. The Muhammadan barbers are somewhat more enterprising, and practise lithotomy.

* This arrangement is at present (1925) under suspension, during the absence of Dr. Macphail in Europe, and the work is being done by the district board.

INDIGENOUS. SYSTEMS OF MEDICINE.

The following account of the methods of the *kabirajes* is reproduced from Sir William Hunter's Statistical Account of Monghyr: "The *kabirajes* or Hindu physicians of Bihar are possessed of a system of medicine which, in the hands of the more educated members of the profession, is on the whole rational, though founded on a vague and hypothetical knowledge. There are besides a large number of quack doctors, in league with the village *ojhas* or spirit charmers and low Brahmans, who recommend incantations, charms, and the performance of *pujas*. In the list of indigenous drugs used in the native pharmacopœia, the number of aphrodisiacs is very striking. It is in this class of medicines, and amongst the people who use them, that the practice of the quacks is largest. At the outset of cholera, *kabirajes* usually administer a pill containing opium, camphor, and nutmeg. In collapse they prescribe *rasun*, a medicine whose action is stimulating; as a diuretic, the fœcal matter of mice moistened with the juice of the plantain tree is applied to the navel. When thirst is excessive, an electuary made of honey and cinnamon powder is placed on the tongue; water is given sparingly. In their medical works eight kinds of fever are described. Of these the principal are the nervous, the bilious, and the catarrhal. The most complicated form is that in which the symptoms of all these three forms are present. The remedy suited for the treatment of all kinds of fever is called *sudarsan churna*, containing fifty medicinal substances, most of which possess febrifuge properties. Purgatives are never given at the outset. Fasting is strictly enjoined during the first four or five days of the attack. In dysentery and diarrhoea, the medicines commonly given are taken from the class of carminatives and antiperiodics. A compound medicine used in fevers contains, amongst other drugs, mercury, gold, talc, copper pyrites, and *haritaki* (*terminalia chebula*). For spleen a compound called *panchan* is used, which consists of the following ingredients :—aloes, lime-juice, rock-salt, black-salt, and vinegar. For dysentery, cloves, *ajawan*, assafoetida; rock salt, black pepper, bay leaves and mint are used in various combinations. Cow's urine is often administered in liver diseases."

CHAPTER V.

AGRICULTURE.

GENERAL CONDITIONS.

AGRICULTURAL conditions are very different in the north and south of the district. To the north is a fertile alluvial plain devoid of hills or natural eminences. In the west of this plain, from the boundary of the Darbhanga district to the mouth of the Gandak, the land is almost entirely under cultivation, the chief crops being *bhadoi* and *rabi*. The depressed tract to the east grows fine *rabi* crops in some places and paddy in others, but during the rains it is to a great extent inundated and uninhabited, and there are large tracts of pasture land, where herds graze in the dry and hot weather. South of the Ganges the cultivated area lies chiefly in the basin of the Kiul river and its tributaries, and in *pargana* Kharagpur, where the largest area is under winter rice. The tract to the north of Sheikhpura and west of Lakhisarai, which is also liable to inundation, is nearly all devoted to *bhadoi* and *rabi*. The following is a brief account of the different tracts of fertility.

TRACTS OF FERTILITY.

Diaras.

Some of the most fertile lands in the district consist of the *diaras* of the Ganges, i.e., lands in the bed of the river which are constantly being added to or diluviated by floods. The creation of these *diaras*, or *chars*, as they are also called, is an interesting example of soil formation. Some back-water or curve of the river bed sets up an eddy in the current, which thereupon becomes sufficiently stationary to deposit a portion of the sand which it holds in solution. The level of the *diara*, which is so far nothing but a heap of sand, then gradually rises as the water lying stagnant spreads a thin layer of clay and silt over the sand; and this deposit of silt deepens at every high flood, until at last the *diara* rises above flood-level. The soil of such a *diara* is extremely fertile, and grows magnificent crops; but if its growth is arrested by the river altering its course, so that the flood water does not cover it during the second stage of its formation, it remains sandy and barren.

Some *diara* lands are the most fertile in the district, producing fine *bhadoi* crops before the river rises and good *rabi* crops in the cold weather. Other *diaras* again may be all sand, and the good field of one year may be ruined by a deposit of sand in the next. Cultivation on *diaras* is thus often a mere speculation. These lands are also the subject of perpetual dispute and frequent litigation, which is of a complex nature owing to the absence of fixed landmarks, and the difficulty of knowing whether the land is an accretion or a re-formation *in situ*.

North Monghyr.

North of the Ganges, about three-fourths of the Begusarai subdivision lies between the channels of the Burh Gandak and the Ganges, and consists of the older alluvial deposits of those two rivers, well raised above their beds and comparatively immune from injurious inundation. The northern portion of thana Begusarai and almost the whole of thana Gogri, which covers the eastern half of North Monghyr, is a low-lying tract singularly liable to destructive flood. To the north-west is the Kabar Tal, while thana Gogri, which is traversed by the Baghmati and Tiljuga rivers, is lined with a series of *jhils*, the remains of former river channels.

Proceeding from west to east, the Teghra thana consists mainly of uplands suitable for the cultivation of autumn and winter crops. The same description applies to the southern portion of the Begusarai thana; but, in the north-west, the tract north of the Burh Gandak includes a considerable area suitable for the cultivation of winter rice. The centre of this latter area is occupied by the Kabar Tal, a large shallow lake, extending over nearly seven square miles, a portion of which is always under water and uncultivable. The remainder dries up in time to allow the sowing of rice broadcast in the month of May, the crop being reaped from boats in November. East of this lake the country becomes more like that characteristic of thana Gogri, with stretches of grass jungle interspersed with rivers and *jhils*. In the last few years much grass jungle has been replaced by *rabi* crops; and it is probable that eventually the larger part of the area will be reclaimed. Thana Gogri includes a fairly extensive block of high land in the north, where excellent *bhadoi* and *rabi* crops are raised. The central portion of the thana is occupied by swamps and grass jungle, but the latter is in process of reclamation. A large block between the railway embankment and the Ganges is regular Gangetic *diara* land, in which, in favourable years, rich crops of maize are reaped, and in other years fine crops of pulses, wheat, gram, and other *rabi* cereals. In the last few years a noticeable change has been effected in the north-east of this thana, where the crops used to suffer year after year from the overflow of the Kasraiya river. This tract is now safe from inundation owing to the construction of an embankment at the mouth of the river by the Bengal and North-Western Railway; and the land yields equally good *bhadoi* and *rabi* crops.

South Monghyr.

South of the Ganges there is a quasi-*diara* tract along the bank of the Ganges, which bears rich *rabi* crops; while between the river and the East Indian Railway loop line, from Jamalpur to Lakhisarai, there are excellent rice lands, which yield exceptional crops in seasons free from floods. To the north-west, the portion of the Sheikhpura thana lying between the South Bihar and East Indian Railway lines comprises two fairly distinct tracts. The eastern portion is liable to inundation from the Halahar river and has heavy soil growing good *rabi* crops. The western portion, which is irrigated but not flooded by the Sakri river, produces winter rice, the *rabi* crop being comparatively insignificant. South of the railway line from Bariarpur to Sheikhpura, we find three distinct areas. To the extreme east is the valley of the Man, which is irrigated from the Kharagpur reservoir and is mainly under rice. In the centre are the Kharagpur Hills, a mass of rock and jungle with occasional patches of cultivation in the valleys. In the west the tract extending from the Kharagpur Hills as far as the western boundary, which is comprised in the Sikandra thana and the southern portion of the Sheikhpura thana, is a wide flat plain with a slight fall from south to north. It is mainly cultivated with winter rice, but *rabi* crops are also grown to a considerable extent. The soil is dry, and the rice crop is apt to fail from insufficient rainfall.

The southern portion of this latter tract contains chiefly paddy land of a quality much inferior to the alluvial lands nearer the Ganges. The rice crop is entirely dependent on the rainfall, for the scanty provision for water storage cannot compensate adquately for the rapid drainage of the country. Consequently, short monsoon rainfall, followed by a failure of the rain in November and the beginning of December, involves a crop of half or less than half the normal yield. Round the village sites in this tract, the land, manured by generations of good Hindu cultivators, yields mustard, maize, *janera*, castor, *rahar* and sugarcane, and occasionally barley and wheat. Irrigation is effected by means of wells, generally mere pits unlined with masonry, of which there may be several in one village, each owned and kept up by a combination of raiyats, who take the water by turns according to their needs. On the higher lands of the village, out of reach of artificial irrigation, are grown scanty crops of *kulthi*, *kodo*, *til*, mustard, *sarguja*, *rahar* and other oil-seeds and pulses. These higher lands, generally known as *tanr*, are composed largely of sand and gravel with a scanty admixture of clay, and are only moderately fertile, producing a yearly crop with difficulty.

The last division consists of the undulating country stretching from the Kharagpur Hills southwards to the border of the district. In the Jamui thana, in the north of this tract, there are extensive areas under rice, and in the extreme south the cultivators carve out paddy fields by levelling the beds of streamlets, and terracing the sides of the wider and shallower valleys. On the banks of such streams as retain a small amount of water during the dry season, some few sugarcane fields may also be found and an occasional crop of barley or wheat. The greater portion of the land, however, cannot be irrigated. The hard and stony sides of hillocks, the still harder mounds of *kankar* in pargana Parbatpara and *taluks* Mahapur and Dumri, and the flat or hummocky laterite deposits of Chakai, are scratched with a plough once every three years and produce a small crop of *kulthi*, *kodo*, *til*, mustard or *rahar*. The lands nearer the villages, to which manure, in the form of cow-dung and wood ashes, can be conveniently applied, are more fertile, yielding fair crops of barley, castor, *janera* and maize. But these also are mainly dependent on a good rainfall for their full growth, and absence of winter rain is fatal. Every year sees the improvement of cultivation by the conversion of the lower *tanr* lands into paddy fields, where the bed of a stream affords opportunity for irrigation. Such extension is due entirely to the energy of the raiyats; and there is a complete absence of the development of an intelligent system of irrigation works, which can only be carried out by means of the capital of proprietors.

IRRIGATION. Artificial irrigation is little practised or needed in North Monghyr, where the country is subject to inundation during the rains. Irrigation is not resorted to at all in low-lying rice tracts; and statistics compiled during the settlement operations showed that only $2\frac{1}{2}$ per cent. of the total cultivated area was irrigated. Wells are used very little for irrigation except in the Teghra thana, where valuable crops, such as chillies and tobacco, are grown on small patches which can easily be watered from adjoining wells. A small amount of irrigation is effected by means of tanks, but more by small channels and by lifting water from rivers and lakes. In Gogri thana the latter furnish the main source of irrigation for the rice crop, which is more largely cultivated than in the two western thanas. Even here, however, irrigation is so little resorted to that barely one acre in every ten under rice is irrigated, and these figures are hardly representative of the whole thana, as it is only in a small area in the north that irrigation is extensively resorted to.

In South Monghyr irrigation is practised far more freely, affecting 42 per cent. of the net cropped area. The irrigation is of three main kinds, viz., from artificial water channels called *pains*, from artificial reservoirs called *ahars*, and from wells. There is also a system of distributaries leading off from a reservoir at Kharagpur to the north-east, which will be dealt with later. Well irrigation is largely used for sugarcane and vegetables, especially potatoes. There are numbers of *pakka* or masonry wells, and every year a large number of *kachcha* or earthen wells are dug, which will last for a few seasons.

Pains are artificial channels leading off from a river or stream or from an *ahar*, but as a rule the latter is more frequently an independent source of irrigation rather than the storage area for *pains*. These water channels are much in evidence in the west of the sadr subdivision, in the thanas of Surajgarha, Lakhisarai, and Shaikhpura, where the proportion of the net cropped area which is thus irrigated is respectively 46, 26, and 37 per cent. When the *pain* is to be filled, an earthen dam is constructed, which diverts the water into the *pain*. The water is ordinarily carried into the fields by means of the artificial lifting apparatus described below. In order to facilitate the lifting operations by providing deep water [illegible] the place where the lifting is done, dams are constructed [illegible] the *pains*. Each river or stream ordinarily serves several *pains*; and when rainfall is short, the upper dam may cut off the water which should fill *pains* lower down. The utilization of the water of the upper *pain* is necessarily delayed by the fact that the flow of water has to be held up at each lifting station; and if water is scarce the cultivators who benefit by the upper *pain* are naturally unwilling to cut the dam in the stream, though villages below may be in need of water. *Pains.*

The systems are worked generally in a spirit of give and take; they give a considerable amount of work to the courts, civil and criminal. The criminal case arises out of the action of riotous assemblies applying the principle of self-help, often with fatal results to one or more members of one or both of the contending parties. When it is remembered how vital the necessity of water may be, and how much the cultivators of land watered from higher reaches stand to lose by giving up any of the advantages of their favourable position, it may be recognized that there is no ground for surprise that riots occur, but rather ground for admiration of the spirit of mutual forbearance which makes them comparatively infrequent.

Ahars.

Ahars form the principal source of irrigation in the Jamui subdivision. The percentage of the net cropped area in South Monghyr irrigated from these reservoirs is as follows :—

Thana	Monghyr	...	...	22
	Jamalpur	...	...	24
	Kharagpur	...	...	6
	Surajgarha	...	...	45
	Lakhisarai	...	...	53
	Shaikhpura	...	...	35
	Sikandra	...	...	79
	Jamui	...	...	63
	Chakai	...	...	54

An *ahar* is an U-shaped or rectangular tank, which is supplied with water by a *pain*, or by an artificial catchment basin placed across the line of drainage. Embankments are built on three sides of the rectangle, the highest bank being at the end where the water would ordinarily emerge, while one side is left open to allow the water to enter. If a small *ahar* is built across a drainage channel, a narrow cut is made at the deepest end to let out surplus water; if the *ahar* is a large one a weir is made for this purpose, so that the water may escape and fill other *ahars* lower down. The water flows through a weir from the *ahar* to the channels leading to the field; when the water is low it is taken from the *ahar* by means of the lifting arrangements described below.

Lifting arrangements.*
Latha.

The *latha* is a long beam working on an upright forked post which serves as a fulcrum. The beam is weighted at one end with a log or stone, and a cone-shaped bucket (*kunri*) is

*** Pictures illustrating these devices are given in Dr. Grierson's 'Bihar Peasant Life'; *latha* at page 206, and *karin* and *chanr* at page 210. Reference to these pictures will make the descriptions plain.**

attached by a rope to the other end. The cultivator pulls down the rope till the bucket is immersed, the weight attached to the lever then lifts it, and the bucket is emptied into the water channel.

The *karin* is a water scoop shaped like a "dug-out" canoe cut in half. It is usually made of a single piece of wood, but iron *karins* are by no means uncommon. The broad open end rests on the water channel which is to irrigate the fields, and the pointed end is dipped into the reservoir. The water is raised by a lever overhead with a weight at the end of it. The *karin* is used for raising water from *ahars* or from a lower channel to a higher, where water is plentiful, and has not to be lifted to a considerable height. *Karin.*

Where the level of the water in a stream or tank is very little below the level of the land to be irrigated, the water is often raised by means of a *chanr* or *sair*. This consists of a piece of closely woven bamboo matting, about 18 inches to 2 feet square, with two of its corners brought together and sewn up. Two ropes are attached to the wedge-shaped end so formed, and one to each of the pieces of wood fastened across the mouth to keep it open. Two men, standing one on each side of the pool of water, dip the *chanr* into the pool, swing it up to the bank, and tip the water into the channel by sharply raising the ropes. Water is sometimes raised to a high level by means of a series of these *chanrs*, when the fields to be irrigated form a succession of terraces on a slope. *Chanr.*

Little irrigation is possible in the hilly region of the Chakai thana without an outlay of capital, and few wells exist or can be dug in this area. But a good deal of rice is grown in the beds of hill streamlets, which the skill of the Santals turns into fertile fields. An embankment is placed across the stream near its source, and its bed beneath the embankment is levelled into fields, which continue one below the other, like the steps of a stairway, down the whole length of the stream, till it reaches the main stream or strikes soil which cannot be cultivated. The embankment retains the water of the stream till long after the paddy crops are ripe, when otherwise it would have flowed off in a few hours. Each field, too, acts as a small reservoir for the field below it; and all the fields are thus assured of a continuous supply of moisture and, except in extremely dry years, produce a good crop.

Some of the reservoirs thus formed are very large and supply an extensive area with water, but a good deal of cutting is required in order to provide channels leading from the bed of the stream in which the water originally collected. This cannot be done by the raiyats, as a considerable outlay is involved; and unfortunately most landlords of the Jamui subdivision take little interest in the construction or maintenance of irrigation works. In the alluvial plain also, the tanks and *ahars* constructed generations ago, when proprietors seemed to take a more active interest in the welfare of their tenants, or at least in the justifiable increase of their rent-rolls, are also silting up and falling into disrepair. This is regrettable, for the land is in general incapable of producing good crops without irrigation.

Kharagpur reservoir.

The only large irrigation work in the district consists of a reservoir in the Kharagpur estates of the Maharaja of Darbhanga, which was constructed about fifty years ago, when these estates were under the management of the Court of Wards. About two miles south-west of Kharagpur the river Man runs through a gorge between two steep hills, and at the narrowest point of the gorge there is an outcrop of rock across the river bed. This was taken advantage of as a natural foundation on which to raise a great dam. Below this, the gorge widens out into a valley hemmed in by low but abrupt hills, which forms the reservoir. The work was begun in 1870 and completed in 1877, the expenditure being Rs. 6,84,916.

The dam is a mass of earthwork 81 feet high, measured from the river bed, and 27 feet wide on the top, the extreme length on the top, from hillside to hillside, being 700 feet. It is composed of clay rammed in thin layers, and is traversed in the centre by a rubble stone masonry wall, three feet thick at the top and ten feet wide at the base, spread out by footings to 26 feet in the foundation, which rests on solid clay. The front of the slope is protected by a layer of rubble stone about a foot thick. A waste weir, 100 feet long and 340 feet wide on the crest, has been cut out on the south side from the rock, the greatest depth of cutting being forty feet. There are outlet irrigation sluices on the north and south of the river, which have a cut stone floor and rubble stone apron, extending for 200 feet. The walls are faced with cut stone, and pierced in the case of the northern outlet by three vents of 4 feet by 3 feet, and in the southern outlet by four vents of the same size.

They are fitted with iron frames and gates, worked from above by powerful screw gearing. Besides the waste weir, some waste sluices of similar construction are provided, in order to return to the Man river below a certain proportion of the water.

The following account of the capacity of the reservoir is condensed from the *Statistical Account of Bengal*: "The area of the catchment basin is 36 square miles, and the area of the reservoir at the level of the hills is 1·7 square miles, calculated from a survey and a contour section of the basin. The capacity of the reservoir storage between the sills of the sluices and the crest of the waste weir is 1,090,045,440 cubic feet, which represents the amount of water to be impounded in one year for the purpose of irrigation. Assuming a 60 inches rainfall, the total catchment of the basin will be 4,516,300,800 cubic feet, i.e., the sluices and waste weir must be able, if required, to discharge 3,426 millions of cubic feet. These figures are thus fixed in consequence of an agreement between the Darbhanga Court of Wards and Raja Lilanand Singh, who owns the riparian land further down the river, that not more than one-third of the average amount of water due to rainfall should be impounded. For irrigation purposes, there is a storage of 1,090 million cubic feet, or 124,620 cubic feet for each acre of the area to be irrigated, exclusive of the local rainfall over that area." The above calculations were based on the area of the catchment basin, the height of the crest of the waste weir from the sills of the sluices, and the estimated rainfall. But the height of the crest of the waste weir was subsequently reduced; a rainfall of 60 inches is far above the average; and it is stated that the calculations did not allow for the influx of water from the springs of Bhimbandh and Lakshmikund, which are estimated to supply 1,974,720 cubic feet in winter and 987,360 cubic feet in the hot weather.

The area of the land to be irrigated was fixed originally at 26,240 standard *bighas*, but ten years later the area under irrigation from the water of the reservoir and the springs was 47,500 *bighas*, after allowing for the share of the Banaili Raj according to the agreement. It is reported that water is now distributed by means of irrigation channels over about 18,000 acres (approximately 54,000 *bighas*). The scheme has helped to raise the rent-roll of the Darbhanga Raj estates from about Rs. 40,000 to nearly Rs. 1,30,000. The chief irrigation channels consist of the Rahmatpur, Parsanda and

Bhusichak canals on the north of the river Man, and the Muzaffarganj, Kathutia and Buhara branch canals on the south. These canals and distributaries are over thirty miles long.

Gilandazi bandhs.

Outside the area irrigated by the canals and distributaries in the Kharagpur *pargana*, there is a system of irrigation by means of *gilandazi bandhs* or *dhar bandhs*, embankments across hill streams. In this *pargana* it has long been the custom to construct such embankments in order to intercept water for purposes of irrigation. They were formerly constructed, at the expense of the estate, by the *jeth-raiyats* (headmen) of the villages; but as they were made without the necessary careful calculation of the amount of flood-water to which they were liable to be exposed, they were constantly being injured. It was therefore proposed, in 1873, that a special examination should be made of all the embankments, and that they should be divided into three classes :—(1) Those likely to prove permanently remunerative, owing to the increased value given to a large area of land by their construction; (2) those not likely to prove remunerative, but which having been always maintained at the expense of the estate could not be abandoned without hardship to the raiyats, and which could be constructed in a solid manner at an expense not greater than the amount which, at five per cent., would yield the average annual expenditure on repairs; (3) similar *bandhs* which should be maintained, but which could not be constructed within the above limit of expense. It was also suggested that it might be found desirable to erect *bandhs* at places where they had not hitherto existed. Sanction was given to an expenditure of Rs. 30,000, and the construction of *bandhs* irrigating 16,000 *bighas* was completed in 1877.

SOILS.

The soils of the district are (1) heavy clay called *karail*, (2) clay called *kewal*, *kariya* or *kathauk*, (3) clayey loam called *dhusi kewal* or *phulauk*, (4) loam called *dhus*, *dhusri*, *doras* or *balmat*, (5) sandy loam called *balsumbhri* and (6) sand or *bal*. The clay soils of the district are classified as follows :—(*a*) *Kachhua kewal* is the typical clay soil of *chaur* or low lands, which remain too long and too deep under water to admit of paddy cultivation. The soil, however, grows all sorts of *rabi* crops. Its colour is black. (*b*) *Karail* or *karari* is a black, tenacious rich soil, also found in the *chaurs*, which grows only *rabi* crops. (*c*) *Dhusri* or *dhusri kewal* is a little lighter than *kachhua kewal* and grows both paddy and winter crops, (*d*) *Gorki* is an extremely stiff soil suitable neither for paddy nor wheat or barley, which grows only *rahar*, gram, *kulthi*, etc. It has a mixed white and red colour. (*e*) A brick-red soil found near the hills, very stiff and impervious to water, rather poor, growing only *rahar*, gram, *kulti*, etc., has no special name.

The usual loamy soil is *bhusni* or *dhusri* or *dhus*, a light rich soil suitable for crops. A sandy soil is known as *balsumi* or *balsumbhri*. It is not a rich soil, but will grow both *bhadoi* and *rabi* crops. *Dhus*, *balmat* or *doras* is a *diara* soil containing about three-fourths sand and one-fourth clay, which yields only inferior *rabi* crops. Alkaline soils are known as *nonchhal* or *usar*, when impregnated with saltpetre (potassium nitrate); as *reh*, when impregnated with sodium carbonate, used by washermen for washing clothes; and as *kharwa*, when apparently containing sodium sulphate. But there is a good deal of confusion about these names.

In North Monghyr an ample and well-distributed rainfall is not so vitally important as in South Monghyr, because the area under rice is comparatively small. If the rainfall is short, lands that would not otherwise be cultivated are available, especially for *rabi* crops; while *diaras* and other lands which are usually flooded yield bumper crops of maize, which in other years might very possibly have been swept away or destroyed by floods. In South Monghyr, on the other hand, the main staple is winter rice; a large proportion of the area under cultivation consists of a dry soil immune from inundation; and here artificial irrigation is essential to guard against the effects of short or unseasonable rainfall. **RAINFALL.**

The net cropped area is 1,119,523 acres, of which 675,458 acres are in North Monghyr, and 444,065 in South Monghyr. **CROPS.**

The following table shows the area under *bhadoi*, *aghani*, and *rabi* crops respectively :—

	Bhadoi.	*Aghani.*	*Rabi.*	Twice-cropped.	Net cropped area.
	Acres.	Acres.	Acres.	Acres.	Acres.
North Monghyr ...	288,682	198,094	444,707	256,025	675,458
South Monghyr ...	148,545	428,968	407,494	190,994	794,013
Total ...	437,227	627,062	852,201	447,019	1,469,471

Rice.

The most important cereal is rice, especially *aghani* or winter rice, which is the main staple south of the Ganges (120,040 acres in North Monghyr and 320,000 acres in South Monghyr). In North Monghyr winter rice is grown chiefly in the Gogri thana, where it occupies 25 per cent. of the net cropped area. Much of it consists, however, of the coarser varieties, which are sown broadcast on the edges of *jhils* and swamps; though they give little trouble to the cultivator, they are less prolific and produce a less valuable grain than varieties cultivated elsewhere with greater care and at greater expense.

Bhadoi rice, a comparatively unimportant crop, is grown on 19,000 acres in North Monghyr, and on 18,000 in South Monghyr. This is also known as *sathi* rice, because the period between sowing and reaping is about sixty days.

Maize.

Maize (232,118 acres) covers the largest area after rice; it covers a larger area than rice in North Monghyr, where it is grown on 139,857 acres. It is regularly sown in *diara* areas on the chance that the flood may come sufficiently late to allow of the crop being reaped, and also in low-lying land liable to flood from the Ganges or Gandak.

Gram.

Gram is a crop of considerable importance in South Monghyr, where it is grown on 141,000 acres, a staple *rabi* crop in all parts of this area except the thanas of Jamui and Chakai. In North Monghyr, where it is grown on the *diaras*, besides being sown as a second crop after rice, it is cultivated on 68,700 acres.

Wheat.

Wheat (145,000 acres) is cultivated largely in North Monghyr, where it is grown on 99,400 acres. This delicate cereal is most extensively grown in Begusarai thana on lands on which the annual inundation of the Ganges leaves a rich deposit of silt. In such areas the ordinarily heavy cost of cultivation of this crop is to a large extent avoided, and the cultivator is thus able to bear with comparative equanimity the chances of loss through blight, to which this crop is particularly liable.

Barley.

Barley (81,000 acres) is mainly cultivated in North Monghyr, where it covers 59,500 acres. In the poorer lands of Gogri it to some extent takes the place of the more valuable spring crops which can be raised in other thanas.

Marua.

Marua is, next to maize, the most important *bhadoi* crop and is principally grown in North Monghyr.

Other cereals and pulses. The most important of the remaining cereals and pulses are *arhar*, *khesari*, *masuri*, peas, oats, *kodo*, *sama* (*sawan*), *china* and *kulthi*, which are grown over a very large area in North Monghyr. Of these, *khesari*, peas, and *kulthi* (vulg *kurthi*) are grown as second or catch-crops after the harvesting of the winter rice, particularly in thana Gogri. The *china* millet is sown late in the cold weather, especially in the area south-west and west of Sheikhpura. It is not of much importance, but it strikes the eye with its fresh green, when the main *rabi* crop is already ripe, if not harvested. It depends on irrigation, chiefly from wells.

Oil-seeds. Oil-seeds are the main non-food crop of the district and consist chiefly of linseed (37,500 acres) and rape and mustard (29,000 acres). In the cold weather the fields in the *rabi*-growing tracts are yellow with the crop last named. *Til*, or gingelly, is raised in the south near the hills; and the remaining area under oil-seeds is given up to miscellaneous crops, such as castor-oil.

Sugarcane. Sugarcane is a crop of little importance in North Monghyr, where only 1,300 acres are planted with it. It is cultivated principally in South Monghyr, where it is of some importance to the north-west and also round Sikandra. In South Monghyr the area under this crop is 8,500 acres.

Fibres. Fibres are not cultivated in Monghyr district; cotton is grown on 400 acres, and other fibre crops scarcely at all.

Indigo. Indigo was formerly an important crop in the Begusarai subdivision. In 1901-2, in the course of the preparation of the record-of-rights for North Monghyr, it was found that 15,026 acres, or 4½ per cent. of the cultivated area of the Begusarai subdivision, were cultivated with indigo by or on behalf of factories in the Teghra and Begusarai thanas. But even then planters were beginning to substitute sugarcane and other country crops for indigo; and by 1908 the area under indigo had shrunk to 6,400 acres. In 1924 the area under the plant was 1,800 acres.

Tobacco. Tobacco is a crop of some importance in the Teghra thana, where nearly three per cent. of the net cropped area is devoted to it. In South Monghyr it is grown here and there close to the village sites, but not on any large scale.

EXTENSION OF CULTIVATION. The normal cultivated area is sixty per cent. of the total area of the district; but the possibility of any considerable extension of cultivation is limited by the fact that south of the

Ganges a large tract consists of hills, while north of the river there is a large area of swamp in Gogri thana. The area returned as not available for cultivation is 761 square miles, and the area of culturable waste is 695 square miles. There is ample evidence of the extension of cultivation both in the north and south of the district. Fifty years ago the Collector reported—" There appears little doubt that so great is the demand for land, that a good deal of land, which ten years ago was considered not worth cultivation, has during the past two or three years been broken up. I may mention the broad belt of land adjoining the Kharagpur Hills west of the Kharagpur road, and a very considerable area to the north-west of Pharkiya and throughout the central parts of that *pargana*. The jungle lands, conspicuous in the survey maps in Pharkiya, are now very generally studded with fields of corn." The advance made in *pargana* Pharkiya or thana Gogri during the eighty years which have elapsed since the revenue survey may be gathered from the fact that only 35 per cent. of its area was then cultivated, while now 64 per cent. is under cultivation. Much of the land under tillage has only recently been reclaimed, and more might be reclaimed were it not for the liability of this tract to flood, owing to which cultivation in the swampy areas is still carried on to a large extent by non-resident cultivators called *dohatwars* or *pahikashtkars*.

In South Monghyr the construction of the Kharagpur reservoir has resulted in a considerable extension of cultivation in the north-east, and in the north-west practically all culturable land has been taken up except at the foot of the hills. Much of the waste in this latter tract has been cleared within the last eighty years to judge from the revenue survey maps, but the process of clearing has now apparently well-nigh reached its limits. At present cultivation in the neighbourhood of the hills is fluctuating, lands being tilled one year and abandoned the next.

In the south there is a good deal of jungle and waste land awaiting the woodman's axe and the cultivator's plough, but extension of cultivation is practically impossible without extension of irrigation. There are signs that tracts of *tanr* land, previously cleared of timber to grow *kulthi* and other pulses, have been again allowed to relapse into waste, and are now covered with thorny scrub jungle. In many places these are being cleared for a second time, and paddy fields are laid out in the beds and on the banks of the streams which traverse them. The Santals of Mahapur Kalan and Chakai are gradually extending cultivation in their villages by the system of damming up streams already mentioned; and every year sees the further retreat of the fringe of jungle towards the higher hills. Where, however, irrigation is not possible even in this primitive form, the clearance of jungle is not synonymous with extension of cultivation. Large areas are fit for nothing else but the growth of forest and poor forest at that. A few crops of *kodo* and *kulthi* may be reaped from land thus cleared, but without manure and water even this scanty crop ceases to be remunerative; and the thin covering of soil, loosened by the plough, is washed by the rain into the beds of the hill streams, leaving a stony surface and rocks too bare to give a hold to even the smallest timber. As an instance of this process may be mentioned Dhanwe, a large *taluk* of *pargana* Chakai, where the forest has long ago been cleared, and even the paddy fields are losing their fertility. It is made up of hummocky hills, and the chief component of the soil is quartz stone. The stony ground, having yielded a crop or two of *kulthi* to the first clearers, refuses to bear trees, much less crops or grass. The result is that the number of cattle has yearly decreased owing to the absence of fodder, and the decrease of cattle implies decrease of manure, which must be applied in large quantities to terraced paddy fields to make up for the loss caused by extremely rapid drainage. Thus the village, though cleared, has a diminishing rather than an increasing productiveness.

IMPROVEMENTS IN CULTIVATION.

With the exception of iron roller sugarcane mills, improved implements have not come into favour. Formerly the cane was squeezed by being passed between two revolving wooden cylinders, but the pressure thus obtained was weak and uneven, and the operation had to be repeated several times and even then the juice was not wholly extracted. This archaic mill has been superseded by one of iron, by means of which not only is the work done much more expeditiously, but far less juice is left in the cane. The new machine has come into use almost everywhere, and the raucous creaking of the old wooden mills, once so characteristic of the early spring, is now rarely to be heard. The value of catch-crops is well understood, i.e., of crops taken off the lands between the crops of an ordinary rotation when otherwise the land would remain fallow. As an example of such catch-crops may be mentioned *khesari*, which is sown broadcast in the rice fields as they are drying up at the end of the rains. In this case a double advantage is reaped; there is the additional crop of straw and grain, and *khesari* being a leguminous crop, increases the store of nitrogen, i.e., it adds some plant-food to the soil. Gram is similarly sown in rice fields after the paddy has been harvested. With the exception of rice, few of the crops are grown singly. As many as five or six different crops may be seen growing mixed up together in the same field, such as wheat or barely with linseed, mustard, gram, *khesari* and *masuri*.

FRUITS AND VEGETABLES.

The mango is found all over the district, but it is more common north of the Ganges and along its southern bank than in the hill tracts, where it gives place to *mahua* (*bassia latifolia*). The mango yield is very uncertain, but in good seasons it forms a considerable portion of the food of the people in May and June. The *mahua* tree and its products have been already described in Chapter I. The jack-fruit or *kathal* (*artocarpus integrifolia*) is common, and the plantain is found all over the district, being especially common in North Monghyr, but the fruit is usually of a very coarse description. The palm-tree (*borassus flabellifer*) is very common, and produces vast quantities of *tari* (toddy), the fermented sap of which is obtained from the peduncles cut before flowering. The date-palm or *khajur* is also cultivated for the *tari* it yields. The *tamarind* or *imli*, the *bair* (*zizyphus jujuba*), the *sharifa* or custard apple, and the *jam* (*eugenia jambolana*) are numerous, and their fruit is in considerable demand. Among figs, the fruit of the *dumar* (*ficus carica*) and *gular* (*f. glomerata*) are eaten by the lower classes. *Citrus acida* and other species of lime, when properly cultivated, grow to perfection. The wood-apple or *kathbel* and the *bel* (*Ægle marmelos*) are highly esteemed, and are found all over the district. The *karaunda* (*carissa carandas*) is another fruit tree very generally cultivated; its fruit is used to make a jelly very like red-currant jelly, and other preserves. The pine-apple grows only in the north of the district. The white and black mulberry yield fruit in profusion, and the *litchi* (*nephelium lichi*) grows well; but the oranges and pomelos grown in the district are of an inferior description, seldom worth eating. Peach and apricot thrive indifferently, and though the pear tree is found in gardens, the fruit is not good. Other common fruits are the loquat (*lukat*), pomegranate, guava, which grows to a large size, and the papaya.

A large number of vegetables are cultivated, among which may be mentioned potatoes, cucumbers, brinjals, chillies or capsicums, cauliflowers, onions, peas, radishes, garlic, etc. The *ramturai*, known to Europeans as "lady's fingers", is grown in almost every garden. The sweet melon or *kharbuja* grows luxuriantly on the sandy *diaras* bordering the Ganges, and the water melon (*tarbuja*) is equally common. The suitability of much of the soil for vegetable gardening may be gathered from an account of a vegetable and flower show held in the Government gardens at Monghyr in 1877. "In order to show what the ground can produce when properly cultivated and manured, the Superintendent, General Murray, exhibited one of the largest groups of vegetables ever seen. They were piled up in a pyramid, and the base, which measured twenty feet in circumference, was trimmed with cauliflowers—each four feet round—potatoes, turnips, carrots, lettuces, cabbages, and green peas, which would have excited admiration even in Covent Garden. Plantains, green coconuts and tomatoes, with capsicums—six inches long—formed a second tier; whilst the whole was crowned with a gigantic species of cucumber—five feet long—which had grown upon the roof of the gardener's hut."*

CATTLE. The cattle are generally small and of poor quality. There is good pasturage among the hills to the south during the rainy season and in the grass lands of Pharkiya in the dry season, but elsewhere grazing lands are deficient, and the account given fifty years ago applies to-day. "Every one who has travelled through this district must have been struck with the miserable condition of the cattle. Their only chance of getting a good meal is by trespassing and eating the growing crops, their usual fodder consisting of the scanty grass which grows along the roadside, and the weeds which spring up among the stubble after the crops have been cut." Even in the south of the district the jungle yields poor grazing during the dry season, and only those who can afford to send them to the north-Gangetic plains in *pargana* Pharkiya keep good stock. There is no attempt at breeding oxen, and buffaloes are under-sized and of poor stamina. Sheep of an ordinary kind, yielding a poor fleece and scraggy mutton, are reared to some extent in Chakai. Horses are scarce, except in the stables of the larger zamindars, and even the ordinary spavined and cow-hocked country ponies are not

* E. Lockwood, *Natural History, Sport and Travel.*

very numerous. The semi-wild pig and the omnivorous goat are the most flourishing domestic animals. The former are kept by Santals and by the lower labouring castes, such as Dosadhs and Musahars, for their own consumption. The latter are kept by the Goalas of every village. They thrive on the jungly herbage which is too scanty for the support of cattle, and are generally of fair size and strong.

Veterinary assistance.

There has been marked progress in veterinary matters since the publication of the first edition of the Gazetteer in 1909. At that time the only hospital in the district was at Monghyr. Now there are hospitals at Begusarai and Jamui in addition, and a dispensary at Khagaria with touring assistants in charge. A touring assistant is also attached to the Monghyr hospital. The district board also contemplates the opening of two more dispensaries in the near future. The hospital at Monghyr treats a large number of patients, 1,215 new cases being admitted in 1923-4. Two thousand nine hundred and sixty-seven patients were treated by the touring staff both at headquarters and on tour during the same year. The ordinary contagious diseases of cattle have not been of a very severe character, but there is a considerable mortality in the low-lying tracts from parasitic diseases. The district board is fully aware of the necessity of enlarging its veterinary establishment.

CHAPTER VI.

NATURAL CALAMITIES.

LIABILITY TO FAMINE.

SINCE the creation of the district, famine has occurred twice, viz., in 1866 and 1874, and there have been two years of scarcity, viz., 1892 and 1897. The experience of these years shows that the south of the district is most liable to famine, because there the people are mainly dependent on the winter rice crop. North Monghyr is almost immune, the only portion affected in 1874 and in 1892 being the north of the Gogri thana, where there is a large area under rice, while in 1897 the whole tract escaped. It would appear that in this area short and even unseasonable rainfall is less disastrous in its effects than in South Monghyr, mainly because the winter rice crop is comparatively unimportant and the tenants are not dependent on a single season's crop but on two. The proverb that it is an ill wind that blows nobody any good is specially applicable to this area. A late monsoon, which elsewhere means the failure of the paddy seedlings, is usually coincident with a late rise of the Ganges; and the result is that the cultivators in the long strip of North Monghyr bordering on the Ganges get a bumper maize crop. On the other hand, an early failure of the monsoon, which involves the drying up of land under winter rice, coincides with an early subsidence of the Ganges floods. The cultivator is consequently able to prepare the soil and sow early *rabi* crops, the money value of which is enhanced by the failure of the rice crop elsewhere, as well as by the fact that they come on the market in advance of other *rabi* crops. In the inland portions of the same area a comparative failure of the early monsoon rains is actually an advantage to the maize crop. Here, too, as in the *diaras*, an early stoppage of the rains gives an early *rabi* season, and as the cultivator relies as much on well irrigation as on rainfall for his most valuable crops, the comparative shortage of moisture is not very material. The following is a brief account of the famines from which the district has suffered.

FAMINES. Famine of 1866.

The famine of 1866 was most severely felt in the south-west and west of the district, in an area of about 1,300 miles, where rice is the staple crop. The crop of 1864 failed to a considerable extent, and, with the certainty of a repeated failure in 1865, the market rate of the commonest sort of rice rose in October of the latter year to 11 seers for the rupee. Other food-grains became proportionately dear, and the inhabitants of the distressed localities began to flock into the town of Monghyr, where the gratuitous distribution of food was begun in November. At the close of the year there was some relaxation of the pressure in consequence of the gathering of the winter rice, and the continuation of public relief was deemed unnecessary. But this crop was also scanty, and after it had been reaped, agricultural labour was again at a discount. The distress occasioned by the prevailing high prices of food now began to be felt in a more severe degree, the rate for rice in March being 10½ to 8 seers, for pulses 15 to 8½ seers and for maize 17 to 13 seers per rupee. In April crowds of paupers frequented the town, and in May the Municipal Committee resolved to undertake some works for the express purpose of assisting the destitute. As the object was to give employment to those who were able to work, but could find no market for their labour elsewhere, the rate of payment was fixed somewhat below the ordinary rates, but no labour was attracted on these terms.

No further relief measures were attempted till the following July, when, in consequence of the distress prevailing, the gratuitous distribution of food was resumed. The price of rice had now risen as high as 7½ to 7 seers, of pulse 10 to 8½ seers, and of maize 14 to 10 seers per rupee. Relief centres were established on the south of the Ganges at Monghyr, Jamui, Parsanda, Sikandra, Sheikhpura and Chakai in July and August and on the north of the Ganges, where distress appeared later, at Begusarai and Teghra in the end of August and September. The month of greatest suffering was September when the price of rice still ranged from 7½ to 7 seers per rupee. In this month the daily average number gratuitously relieved was 2,200; and in October 767 persons were employed on the construction of a new road between Jamui and Chakai. The highest total daily average number of persons obtaining public relief throughout the district at any period of the famine was 3,450. The distress was aggravated by an outbreak of cholera over the whole south-west of the district. Deaths from this disease were very numerous, especially in Sheikhpura and Sikandra, owing to the crowding together of large numbers of people, reduced by want of food to a very low condition of bodily strength. The number of deaths due to disease, assisted or engendered by want, was returned by the police at 605, and deaths from actual starvation at 642.

Famine of 1874.

As in other districts of Bihar, the rainfall of 1871 in Monghyr was above the average, but it does not appear from official reports that the crops suffered. In 1872 the rainfall, normal as far as quantity is concerned, was not happily distributed in the Begusarai and headquarters subdivisions, while in Jamui it was appreciably deficient. The consequence was that nowhere in the district were the crops above the average, while in Jamui they were decidedly below it. On the whole, the seasons preceding the summer of 1873 had not been as prosperous in South Monghyr as in the rest of Bihar; and although there was no actual scarcity, the prices current in the latter half of 1872 and the first half of 1873 show that the food-grain market was from 10 to 20 per cent. dearer than it previously had been. This antecedent period of pressure must have had, to some extent, an exhausting effect on the resources of the people.

The rainfall of 1873 was not greatly deficient in quantity, but it was unequally distributed. In July and August it was in excess; in September, when heavy rain is necessary, it was less than half the normal fall of that month; and in October there was no rain at all. The rainfall therefore was excessive, when excess was likely to be most injurious, and deficient, when a copious downpour was wanted. The consequence of this unseasonable excess and abnormal deficiency was that only three-eighths of the autumn crops were saved; while the outturn of winter rice in the headquarters and Begusarai subdivisions was but one-eighth, and in Jamui subdivision, which is the great rice-producing region of the district, but one-fourth of an average crop. It is, therefore, not surprising that in January 1874 the prices of rice and Indian-corn, the two staple articles of food, were double the normal rates at that time of the year. "It is," wrote Mr. A. P. MacDonnell, "a fact pregnant with meaning that they were dearer than the prices which prevailed in January 1866. In the latter year the high prices of January rose to famine rates in April, and all through the summer and well into the autumn these famine rates prevailed. The result was a mortality which Mr. Cockerell's figures do not attempt to measure, and scenes of ghastly misery, of which I retain a vivid recollection."*

Fortunately, during the earlier months of the year the district food-supply received a large addition from the produce of the *rabi* crops, which yielded three-fourths of an average harvest all over the district. This favourable outturn was doubtless largely due to the fact that much of the soil, being liable to inundation, is to a considerable degree independent of the rainfall. This was not all. The moisture in the soil, which brought the *rabi* to, if not full, at least nearly full maturity, enabled the people, herein far more fortunate than their neighbours to the north, to sow a large crop of subsidiary food-grains, which, the Collector estimated, covered 10 per cent. of the cultivated area of the district. It may therefore be said that the district drew, from internal sources, a supply of food-grain sufficient, had it been freely available, and had none been exported, to have supported the people in their usual state for nine months, or sufficient to have supported them in straitened circumstances over the whole period of the scarcity. But the local food-supply could have done no more than this; it was by no means freely available, and it was largely exported. The consequence was a continuous pressure which, at times and in particular localities, deepened into actual distress, necessitating Government relief.

On the 24th April it was reported that, except in the neighbourhood of Bakhtiyarpur, the condition of the district was on the whole satisfactory. The cold-weather crops had been succeeded bv a good *makua* crop; supplies were ample; and there had been no great export of *rabi* grain, while the number of persons on relief works had fallen to 5,364. At the end of May the condition of affairs was generally favourable, and in most cases improved; but the circle officers of Bakhtiyarpur stated that " there can be no doubt that, but for the timely assistance of Government, many thousands of persons would have died of starvation in this circle ". In the beginning of August none of the subdivisional or circle officers reported any distress; and a large incoming *bhadoi* harvest was spoken of, except on the low lands near Lakhisarai, where this crop was destroyed by floods in the Kiul. On the 3rd October all relief ceased.

---

* *Food-grain Supply* and *Famine Relief in Bihar* and *Bengal*, Calcutta, 1876.

The average daily number of persons employed on relief works was 4,210 in December 1873 and 5,227 in January 1874; it rose to 10,596 in May and to 11,148 in June, and then fell sharply to 3,912 in July, 1,972 in August and 332 in September. The average daily number of persons gratuitously relieved was 3,402 at the end of May and reached the maximum of 5,150 in the beginning of August. It fell to 2,084 in the first part of September and to 100 at the end of that month. Altogether, Rs. 4,03,322 were spent on relief, viz., Rs. 83,865 on charitable relief, Rs. 1,32,993 on wages of labour and Rs. 1,86,464 on loans.

Famine of 1892.

Scarcity occurred in 1892 owing to the scanty outturn of the winter rice crop of 1891, preceded by a short crop in 1890. Anxiety was felt chiefly for the low tract of country in the Monghyr subdivision, on the north of the Ganges, extending to the boundaries of Darbhanga and Bhagalpur; but eventually relief operations were found necessary only in the extreme north of this belt, within the jurisdiction of the Bakhtiyarpur outpost of the Gogri thana. Several influences combined to accentuate the distress in this tract, the extreme poverty of the population in all seasons; the sandy nature of the soil over most of the area; and the failure of the rains for two successive seasons. The tract covers an area of some 400 square miles, of which the western portion is flooded during the rainy season and is devoted to paddy. The almost complete failure of the paddy in 1891, after a very small crop of 1890, rendered the larger part of the population wholly dependent upon *sag* and roots.

Relief works were started in February 1892, but were at first very poorly attended, apparently because of the recollection of the relief operations in 1874. In that year also Bakhtiyarpur was a relief centre, and the raiyats had a vivid recollection of how grain carts rumbled into their villages, and they fared more or less luxuriously; and they fancied they would be treated in a similar manner if they refused to do coolies' work on the roads. The numbers on work continued to be low till the latter part of May, when the excavation of some tanks in the extreme north of the district resulted in a sudden rush of labourers. The highest average daily attendance on relief works (1,815 persons) was reached in the week ending the 2nd July. After that the situation was so far relieved by opportune rain and the prospect of a good *bhadoi* crop, that the number of labourers rapidly dwindled away until the relief works were closed on the 23rd July. Gratuitous relief was commenced in the latter part of February, and continued till the 20th August, i.e., a month longer than the relief works. The average daily number so relieved was 235 for a period of 24 weeks, the daily number seldom falling below 200, but never exceeding 300.

The following summary of the main features of the famine is extracted from the final report of the Collector (Mr. C. A. Oldham, I.C.S.) :—"Relief operations were undertaken on but a very small scale in a small and compact area. Appearances were certainly in favour of expecting that a very large proportion of the population would rush to the works. From the small numbers shown in the rolls, a lesson may be learnt, first, that after one year's total failure and a very small harvest of a previous year, the people could still manage to tide over the distress. The only explanation of this is that they keep a year's stock at least in store, and this was proved to me on more than one occasion in the villages. Secondly, the remarkable extent to which the poorer classes can find means of subsistence from roots and bulbs. The remarkable abundance of two roots, *chilor* and *bisaur*, the former a bulb somewhat like a very small potato growing in low-lying swamps, and the latter the root of an aquatic plant, had a very potent effect in allaying the distress. In every village a crowd of women and children might be seen digging up these roots and drying them for food. I calculated that some 5,000 people were living more or less entirely on these roots, and the very unusual drought had dried up the beds of *chaurs* which, in ordinary years, remain full, and so gave access to a larger supply."

Famine of 1897.

Monghyr was scarcely affected by the famine of 1897, which was so severely felt in the neighbouring district of Darbhanga. The rainfall of 1895-6 had been short, amounting only to 33·64 inches, and the crops had consequently been by no means full; but work and money were plentiful, and the cultivators obtained good prices for their produce. In 1896-7 the district shared in the drought which lasted till May; in the break in the rains from the 20th July to 20th August, which spoiled the hopes of the *bhadoi;* and in the final drought from the 24th September to the 31st December 1896. After that date there was good rain every month, and the weather was particularly favourable to agricultural prospects, though not to the mango and *mahua* crops. Not one in twenty of the mango trees flowered, and the *mahua* blossoms being injured by the storms of March, the produce was from a half to two-thirds of the average. On the other hand, the district was fortunate in receiving no less than 11·26 inches of rain in September 1896, and the total rainfall from April 1896 to April 1897 was 41·80 inches. In the end, the outturn of *bhadoi* was 10½ annas, of winter rice 8 annas and of *rabi* 12 annas, taking 16 annas to represent the average outturn.

In one respect the position was peculiar, viz., that the district produced enough food to support its own population; but the famine in Upper India and the failure of crops in Eastern Bengal drained off the surplus, and high prices prevailed not only for rice but for all food-grains. The people, in fact, suffered owing to high prices caused by exportation rather than to bad local crops. Still, though there was some distress owing to the high level of prices, the pinch was felt only in parts of the Jamui and Monghyr subdivisions. Here, however, the poppy was an excellent crop, enabling the people to pay their rents, while they saved their food-grains; and the Begusarai subdivision fared well, owing to the good crops of tobacco and pepper which are extensively grown there. Nothing like famine ensued, the necessary public relief being afforded by means of a fund raised locally and expended on charitable relief. The District Board, it is true, started a road from Chakai to Nawadih in the Jamui subdivision, as a test relief work, in order to ascertain whether the opening of relief works was necessary; but it did not attract labourers.

LIABILITY TO FLOODS.

Though the northern portion of the district is not, as a rule, liable to suffer much from short and unseasonable fall, it is exposed to flood from the overflowing of the Ganges and its affluents. In Gogri thana, with the exception of a small area in the extreme north, and in the north-eastern portion of thana Begusarai, the whole country is regularly flooded by the overflow of the Baghmati and the Tiljuga rivers and their numerous tributaries. The extent of the flood varies from year to year, but, as no attempt is made to grow any but cold-weather crops in the area liable to flood, the damage done is confined to years when the flood subsides too late to allow such cultivation—and this can seldom or never happen except in comparatively small areas where cultivation is ordinarily precarious. In the central portion of thanas Teghra and Begusarai floods are not, as a rule, caused by the overflow of the Burh Gandak, which flows between high and well-defined banks, and any damage that may be caused thereby to the *bhadoi* crops is counterbalanced by the increased outturn of the succeeding cold-weather crops. In the southern strips, between the Ganges and the railway embankment, there are annual inundations, which vary, however, in extent and in point of time. Much depends on the latter factor; in the floods of September 1904 less damage was caused than in those of August 1894, because the crops had already been partly reaped in the former year, whereas the floods of 1894 occurred a week or ten days before harvesting could be commenced.

At the same time, it must be admitted that there is a general belief that since the building of the railway embankment, the floods invariably do damage, however seasonable they may be. In the first place, the embankment confines the flood-water to a smaller area, which is thus flooded to a greater depth than would otherwise have been the case. Secondly it banks the water up, preventing it from spreading northwards and eastwards gradually as it used to do, without doing any particular harm. Now the banked-up water either drowns the crops or flows in a rapid current eastward, breaching *bandhs* and, by its rush, damaging crops. Apart from this however, even a destructive flood has one compensating advantage in the shape of a rich deposit of silt and a corresponding prolific *rabi* crop in the next cold weather.

Of late years the most serious floods have been those of 1894, 1901, 1904, 1905, 1906, 1916 and 1923. From the following description of the floods of 1904 and 1906 an idea may be gathered of the general nature of the floods to which the north of the district is liable.

Flood of 1904.

The flood of 1904 was due to the abnormal height to which the Ganges rose. It was not quite so high as in 1901, when the gauge at Monghyr read 27·75 feet above zero on the 8th September; but in the latter year the flood rose and fell more rapidly. On the 4th September 1901, the water stood at 24·67 feet, on the 5th it rose to 25·75 feet, on the 6th to 26·67 feet, on the 7th to 27·5 feet, on the 8th to 27·75 feet; but it fell again so rapidly that by the 11th it was down to 24·58 feet and by the 14th September to 17·58 feet only. It was over 25 feet for six days only. In 1904 the river rose above 25 feet on the 14th August, attained 27·25 feet on the 19th, and fell to 26·83 feet on the 20th, to 25·75 feet on the 24th and to 24·58 feet on the 29th. It was thus altogether 15 days above 25 feet. Owing to the duration of this flood, the maximum attained and the existence of the Bengal and North-Western Railway embankment, it was probably one of the severest floods ever experienced in the tracts affected. These tracts were (1) the country round Begusarai to the west, (2) the country round Gogri to the east, and (3) a small tract near Monghyr town. It will be convenient to deal with each of these tracts separately.

The flood in Begusarai was mainly due to the fact that the Gupta Bandh, an embankment about 13 miles long extending from Mahna to Samho, was breached in 10 or 12 places, and the Ganges water poured in from the south and west, flooding the whole area south of the Bengal and North-Western Railway line from two miles west of Tilrath station up to Ballia. To the east of Ballia the country was also flooded, but this occurs nearly every year, and was not due to the bursting of the *bandh*. The area flooded in consequence of the bursting of the *bandh* has been variously estimated at 67 or 100 square miles. The depth of water varied according to the configuration of the country. In some low-lying places it was 8 to 10 feet deep; round village sites it was rarely more than 4 to 5 feet deep; and a large number of villages built on higher sites were not touched. The state of affairs in Begusarai itself may be gathered from the report of the Collector, Mr. C. A. Oldham, I.C.S.—" All communication between the railway station and the town and subdivisional office was cut off by the floods. The water was beyond a man's depth, and a swift current was flowing eastwards. Luckily we spied a boat in the distance, which the police had brought to render assistance in the town, and which we hailed and got into. We first proceeded to some houses where 15 or 20 people were sitting on the roofs, and as they were in danger of being washed away, we took them off in two trips to some high land; and then proceeded through streets to the subdivisional office. Here we found the Subdivisional Officer, Mr. Ward, isolated on the upper storey of his house, with inmates of the local subsidiary jail on his roof, and records from his subordinate offices, thana, etc., in the upper verandahs. The flood had come in so suddenly, that he had scarcely had time to save the stamps and notes in the treasury and the records of his office, thana, jail, etc. When we arrived, the water was 9 inches over the plinth. It continued to rise that night, and to make matters worse for the poor people outside, it rained in torrents. Owing to the current, *kutcha* houses were falling on all sides. I decided to get to the railway line, as communication might be wholly cut off at any moment and I could better organize relief from Monghyr. We got into the boat and tried to reach the station, but could make no headway against the current where it was strongest, and were driven back to the subdivisional office and residence : the office is in the lower storey and the residence in the upper storey of the same building. Next morning, I got to the station by going down with the current in the boat and striking the railway embankment. The water was at this time more than a foot deep over the plinth of the lower storey, and three feet deep around the building."

The flood was the highest on record in the town, the next highest flood there, according to local tradition, having occurred some 30 years before, when the water came up to the subdivisional office, but did not top the plinth. In 1901, when the Ganges reached 28 feet (the record for Monghyr since gauge readings have been taken), the Gupta Bandh was also breached, but the water only just reached the subdivisional office drain, and did not actually touch the building itself. On this occasion the water was about two feet higher than on either of the previous occasions, and, moreover, the damage done to houses was very much greater.

The tract that suffered most was that lying between the Bengal and North-Western Railway line and the Tirhut road. Here the water, having topped and breached the road in its progress northwards, was stopped by the railway embankment, and then turned eastwards, a considerable current setting in between the road and the railway, which are not far apart. No less than 1,125 mud-walled houses are known to have collapsed, but the houses in the *diara*, which are built of wattle and thatch, nearly all withstood the flood. The *bhadoi* crops, consisting chiefly of maize, were seriously damaged. Much fortunately had already been harvested, but of the remainder some was entirely destroyed and in places 50 to 75 per cent. was lost.

In the Gogri tract the flood was due to the bursting of the Gogri embankment, which on the 18th August suddenly settled and the water bursting over it soon made a large breach. Luckily the people in the villages on the east of the embankment had been warned to get their property away and be ready to decamp at a moment's notice, and they had acted on this advice, though very reluctantly, so that no lives were lost. Nine villages were flooded, and though a large number of houses were damaged no person was drowned and no cattle lost. The area flooded in this part of the district was 13 square miles.

In the third tract the area affected comprised only Khagaria and its immediate vicinity; but the loss of grain was much greater than anywhere else. Here the inundation was due to the Ganges flood banking up the waters of the Burh Gandak, and the water entered the town by first overtopping, and then breaching, the embankment of the Khagaria-Sisauni road, which serves as a protection to the town on the west. The breach occurred in the evening of the 7th August, and on the morning of the 18th the water entered the western bazar, completely flooding it by the afternoon of the same day, while by the 19th the eastern extremity of the town was also under water.

Besides the above three tracts, several of the *diaras* in the Ganges were flooded, and in places cut away by the force of the current. On 19th August the Basdeopur Taufir *diara* began to cut away, and next day some 1,200 persons with their cattle and property were safely brought away in boats to the mainland. Harnathpur and Raghunathpur *diaras*, in midstream north of Monghyr, next commenced to be cut away by the current. On the 21st instant 7 men, with 43 head of cattle and other property, were safely brought away on boats from Harnathpur *diara*. On the 22nd instant, news having been received that Harnathpur was in extreme danger, a Deputy Magistrate chartered a steamer and boats, and with the assistance of the police rescued some 95 persons, their property, and 2,100 cattle, leaving no one on the *diara*, which was subsequently swept away. All the inhabitants of Raghunathpur *diara* with their cattle were safely brought away next day, and the people were removed without loss of life from Chaintola on the Zamindari *diara*, which was also being cut away.

Flood of 1906.

In 1906 the floods which annually visit Pharkiya *pargana* and parts of the Begusarai subdivision, and which are caused mainly by the overflow of the Gandak and Tiljuga and by the Ganges spill, were unusually early. About the 14th August the Gandak and Tiljuga began to rise to an abnormal height, and about the same time the Ganges also reached a height of over 24 feet. The Collector, who was out at Chapraon, found on his way back to Khagaria on the 17th that except on very high lands nearly all the villages had suffered and that the *bhadoi* crop was practically destroyed. Only in a few villages had the villagers managed to cut and save some partially ripe *marua* and maize. A large number of people had been rendered homeless and were in great distress, the more so as the outturn of *rabi* had been very poor during the previous season and the *bhadoi* in 1905 had also been nearly a failure. The flood continued to rise till the 19th, remained stationary for four or five days, and began to fall about the 23rd. By this time the whole of the Pharkiya *pargana* north of the railway line, with the exception of a few villages and high lands in the north-eastern corner, had been swamped, and entire villages were under water; while in the Begusarai subdivision the Gandak, with its tributary the Balan, had done considerable damage in a strip of land, about six or seven miles board, along the southern bank of the Gandak.

The area between the Bengal and North-Western Railway line and the Ganges, which is liable to inundation by the floods of that river, suffered much less, for it has partial protection from the Public Works Department embankment near Teghra, the Gupta Bandh in the Begusarai subdivision, and the Gogri Bandh near Gogri; and so long as they did not give way, the only areas liable to flood were those lying outside them, viz., the area between the Gupta and Gogri Bandhs, and the area east of the Gogri Bandh. Fortunately these embankments remained intact and the Ganges did not rise very high; and though crops were lost in places south and east of the Gupta Bandh, the damage does not appear to have been as great as had often been the case in previous years.

Altogether, an area of 1,074 square miles suffered from the floods, the distress being keenest in parts of Ballia thana and in the portion of the Khagaria thana bounded on the east by a line from Khagaria to Sarabjita Ghat, on the north-east by the Tiljuga river up to Mohraghat, on the north-west by the boundary of the district, and on the south-west by a line from Bakhri to Khagaria. South of the Ganges the Government tenants in Binda *diara* and Kutlupur lost their *bhadoi*, and much damage was caused in the area between the river and the East Indian Railway line from Monghyr to Kajra. Loans were given to the distressed in the tracts most severely affected, the Gogri and Khagaria thanas and the Bakhtiyarpur outpost. It was at first intended to fix the average of loans at Re. 1 per *bigha* and to distribute no loan to any raiyat possessing more than 10 *bighas;* but this idea had to be given up owing to the serious nature of the distress prevailing. The average was, therefore, raised to Rs. 2 per *bigha,* and agriculturists having 20 *bighas* were allowed loans. It soon became evident, however, that even this was insufficient; consequently, the average was again raised to Rs. 3 per *bigha,* and the maximum amount which a single individual could take was fixed at Rs. 80. Altogether Rs. 56,000 were distributed as loans.

Floods of 1916 and 1923.

In 1916 there was an unusually high flood of the Ganges in the *diara* lands, while at the same time the rise of the Baghmati and the Burh Gandak flooded the northern part of the Begusarai subdivision. There was considerable loss of cattle and destruction of *kachha* houses, but only one human life was lost. In August of 1923 there was an exceptionally high flood in the Ganges, which occurred during a long break in the rains, when the river had begun to fall, and nobody thought of the possibility of a flood. But in the third week of August, though the river was low in Bihar, its higher reaches were in flood; and between the 18th and the 21st of the month there was the great flood in the Son, of which an account may be found in the Gazetteer of Shahabad district.

The Ganges flood came down to Monghyr district on the 22nd of August, rapidly raising the river to a height only three-quarters of an inch below the highest flood-level known. There was no loss of human life, and very little loss of cattle; but the loss of houses was large, because in addition to the loss of the flimsy huts of the *diara* area, built in expectation of destruction by flood, there were many more substantial houses destroyed where the river rose over what is normally its northern high bank.

EMBANKMENTS.

The most important embankments are those of the Bengal and North-Western Railway, viz., (1) the main line, which runs along the south of the north Gangetic portion of the district, at an interval from the Ganges varying from about 12 miles to less than a mile; and (2) a branch line recently constructed, called the Mansi-Bhaptiahi extension, which runs almost due north from Mansi and divides Gogri thana into two equal portions. Both the main line and its extension run along high embankments, and there has been much controversy regarding their effects on the drainage of North Monghyr and its liability to flood. There is, however, a general consensus of opinion that the embankment of the main line is, on the whole, a blessing to the country, in that it protects the area north of the line from the floods of the Ganges—an opinion which is supported by the evidence collected from raiyats by the settlement officers. This question will be dealt with more fully later.

Mansi-Bhaptiahi railway embankment.

The case of the Mansi-Bhaptiahi extension is more difficult. Its embankment runs straight across the natural drainage line of Gogri thana, and the sufficiency of the waterway provided in it is a matter of life and death to the cultivators. By an unfortunate coincidence, the two first years of its construction were marked by exceptional floods. In 1905, there were heavy local falls of rain, which laid the whole country under water and breached the earth-work of the line in many places. This experience was repeated in a more serious form in 1906, when the local floods were due to Sub-Himalayan rainfall. It was found on enquiry that the allowance of water-way was insufficient, and a considerable increase had to be arranged for.

A small area in the west of the Teghra thana is protected by a Government embankment; and in the south and south-east of Begusarai town there is a large triangular area confined between the Ganges and the railway, the major portion of which is protected by the Gupta Bandh, though the existence of the railway embankment intensifies the effects of flood. The rectangular area in the south of thana Gogri lying between the railway and the river, to the south of the Mansi and Maheshkund stations, is similarly protected to some extent by the Gogri Bandh. An account of these embankments is given below.

Private embankments.

There are also some small private embankments in some villages belonging to Babu Ram Bahadur Singh and in the Government estate of Dhanupra in the extreme north, besides an embankment about four miles long constructed in Amosi village in the Khagaria thana during the year 1903. "These zamindari embankments," as Sir Hugh McPherson wrote in 1907, when he was Director of Land Records, "are a source of much difficulty to the District Officer. The greater portion of thana Gogri must, in my opinion, revert to the uncultivated condition in which it was seventy or one hundred years ago, unless the petty embankments are maintained in such a state of efficiency as will protect the country from injurious flood in ordinary years. They are useless in years of exceptional flood like 1905 or 1906, but they give the cultivators confidence and make cultivation possible in the average year. When maintained at ordinary flood level, they do not, I think, seriously endanger the welfare of adjoining tracts. In the winter of 1907-8, about Rs. 20,000 was taken by the Bahadurpur, Shankarpur and Bakhtiyarpur estates of *pargana* Pharkiya in the shape of land improvement loans, and spent on the repair of estate embankments. The expenditure of this money not only improved the efficiency of the embankments, but also afforded relief to the impoverished cultivators and labourers, who had suffered so severely from the loss of crops and destruction of houses attending the floods of 1905-6."

The following is an account of the principal embankments in the district.

Monghyr Bandh.

In the west of thana Teghra a Government embankment, called the Monghyr Bandh, runs along the left bank of the Baya river from the district boundary to Barauni, and protects a large part of Teghra thana from inundation. Further east its place is to some extent taken by the high road from Teghra to Begusarai.

Gupta Bandh.

South of the latter place is an old zamindari embankment, which was strengthened and repaired after the floods of 1889 by the late Mr. Ashutosh Gupta, who was Subdivisional Officer of Begusarai from 1888 to 1890. It is consequently known as the Gupta Bandh. It is an embankment about 13 miles long, starting from near Garhara and terminating on some high land near Saiyadpur below Samho, a large village south-east of Begusarai. This embankment gives partial protection to a large tract of country to the south of the present Bengal and North-Western Railway embankment, but until recently it has never been effective. In 1891 a proposal was made that it should be raised and strengthened, which was negatived by Government. In 1894 the question of remodelling and maintaining the embankment was again raised and negatived, on the ground that the construction of marginal embankments blocks the spill and tends to raise the flood-level and cause damage elsewhere. On the latter occasion it was decided not to bring it under the Embankment Act, but that the zamindars might keep it in repair. It was accordingly, the custom for the Subdivisional Officer to issue notices upon the zamindars to keep it in repair. This expedient, however, was not altogether successful, for the embankment was not properly maintained, portions being left untouched or imperfectly repaired. Moreover, conditions were made worse, and the danger of allowing it to fall into disrepair accentuated, by the construction of the Bengal and North-Western Railway line to the north. This line was opened to traffic in March 1900.

It was found that its high embankment blocked the natural drainage of the country and prevented flood-water escaping to the north; and to make matters worse, a large number of *tolas* sprang up to the south of it, which were more or less dependent on the imperfect protection afforded by the Gupta Bandh. The villagers were, in fact, living as they would in a protected tract and cultivating the land as if it were properly protected, whereas it was not. Their danger was realized when the embankment was breached in the floods of 1904, the highest flood known in this portion of the Ganges. It was finally decided, in 1907, that Government should take over and maintain the embankment under the Bengal Embankment Act (II B. C. of 1882), and that it should be remodelled and raised 3½ feet above the high flood-level of that year.

Gogri embankment.

Further to the east is a short embankment about 7 miles long, known as the Gogri embankment, running from north-west to south-east at a distance of about half a mile from the bank of the Ganges. It was raised by the District Board a few years ago to protect Gogri and a number of villages situated between the Bengal and North-Western Railway line and the Ganges. An embankment at Gogri has existed from a very long time to protect the town of Gogri and the villages to the east and north, but some years ago the southern portion of it was carried away by the river, and with it the old town of Gogri, as shewn in the original survey maps. In 1899-1900 the remaining portion of the embankment was continued as a high raised road running south-east past Gogri-Jamalpur to a village called Bahadurpur at an average distance of half a mile from the river. The flood of 1901, however, breached the new earth-work in the dip where the embankment crosses what is known as the Ratan or Kauwa kol channel; and it was at the same place that the embankment gave way in 1904.

Effect of the railway embankment.

After the floods of 1904 a special inquiry was made by the late Mr. G. G. Maconchy, Superintending Engineer, Public Works Department, into their causes and the remedial measures required. The following extracts from his report give a brief summary of the facts elicited, which will shew sufficiently the effect produced by the embankment of the Bengal and North-Western Railway. Regarding the tract flooded by the breaching of the Gupta Bandh near Begusarai, Mr. Maconchy wrote:—"The tract, bordering the Ganges as it does, has always been liable to floods. The ground is highest near the river, and the fall of the country is away from the river, inland, in roughly a north-eastern direction. There are no drainage lines leading from inland to the Ganges, though there are numerous spill channels leading away from the Ganges. The tract is low, and contains numerous depressions, in which water lodges. The natural direction of the flood-water which tops the bank is nearly in a north-easterly direction towards the Burh Gandak, which traverses the country north of the railway. It is possible that the flood used not to actually fall into this river directly, as there is a low-lying tract between the river and the present alignment of the railway, which may have received the spill from the Gandak as well as from the Ganges. In any case, however, the discharge used to occur into the tract lying to the north of the railway, and the whole volume probably fell into the Gandak near its mouth or rejoined the Ganges through the low lands in that vicinity. It must be concluded that when the tract is flooded by the Ganges, the natural direction of the floods is towards the lower levels of the Gandak spill or of the back water from the Ganges lower down stream.

"This flow is now blocked by the Tirhut road and the railway, both of which traverse the flooded tract in a direction (roughly) from east to west. The Tirhut road is much older than the railway. It was, when constructed, provided with bridges to pass the water carried by the various spill-channels from the Ganges which it crosses. The waterway provided was, however, insufficient, and the excessive scour which occurred through the bridges endangered their safety, and consequently Government in 1888 sanctioned the blocking up of all the bridges. The road has since then formed an obstruction, lying right across the line of direction of the spill, but it has to be remembered that the road is liable to be overtopped and breached by high floods. When the railway was constructed, no waterway was provided, because the alignment ran just behind that of the existing road, in which there was no waterway. Three or four small sluices were, however, constructed to let off the residual drainage after the subsidence of floods. The railway bank is raised well above high flood level and is not, like the Tirhut road, liable to be overtopped and breached, and consequently forms a much more formidable obstacle to the flood-spill. In the flood of 1904 the spill was effectually stopped by the railway, which remained intact, and the flood was ponded up all along its south side, the depth above ground level being about 8 feet at Begusarai, while the ground on the north side was dry. This ponding up shews beyond question that the railway is a complete obstruction to the floods.

"It may be concluded (*a*) that great damage occurred, even before the railway was built, owing to the obstruction caused by the Tirhut road, though it is hardly safe to conclude, again, that the road was responsible for the whole of this damage; (*b*) that the road only caused obstruction in places, and that some outlet was afforded by the overtopping and breaching of the road; (*c*) that the construction of the railway has effectually blocked all outlet for the spill, and rendered absolute the partial evils caused by the road. The higher the flood, the worse comparatively is the effect of the railway, for a low flood might be completely blocked by the road, whereas it inevitably breaches in a very high flood, while the railway remains intact. There are two ways of preventing damage by future floods, viz., (1) to provide sufficient water-way under the railway to allow of the floods escaping towards the Burh Gandak. This would, in a high flood, involve the breaching of the Tirhut road, unless similar water-way were provided in it also; and (2) to build up the Gupta Bandh to a safe height above flood level and of proper section and thus exclude the floods altogether.

"There remains an area of about 10 square miles from which the escape of the Ganges floods towards the Gandak is prevented by the road and the railway. The damage is less severe here, as the water comes in quietly from the Ganges as the flood rises. Still there is some obstruction to the natural direction of the spill, and some water-way appears to be required."

Regarding the other tracts affected by the floods of 1904, Mr. Maconchy wrote :—"The flood at Khagaria was caused by the Ganges flood backing up the Burh Gandak and is easily preventible by strengthening the first mile of the Khagaria-Sisauni road and constructing a cross embankment up to the railway line. The flooding of Khagaria does not appear to have been affected by the closing of the Kasaiya Dhar, as that channel is silted up, and the construction of the railway can only have blocked a small amount of spill through this channel. The line, however, runs in a direction which apparently must shut off a certain amount of spill; and if this is the case, some raising of the general flood level must have occurred. Lower down, coming to the tract flooded by the breaching of the Gogri Bandh, there does not appear to have been any blocking of spill by the railway. The embankment is an old one, and has been breached before. The bank of the river, near which the embankment runs, has cut away a great deal in past years, but this is now said to be stopping. Any checking of flood-spill might tend to increase the cutting again."

As a result of this inquiry, it was decided that the best measure to take was to make the Gupta Bandh effective as a flood bank, by renewing and raising it. and for Government to maintain it under the Embankment Act at the cost of the persons benefited. As a subsidiary measure, arrangements have been made for opening a sluice to a culvert on the District Board road near the Tilrath railway station and for providing a sluiced culvert under the road there.

The inquiry of 1904 dealt mainly with the protection of the Begusarai subdivision. Subsequently, (in 1906) another inquiry was made by Mr. B. K. Finnimore, Superintending Engineer, Public Works Department, regarding the liability of other tracts to inundation. As regards the floods of the Ganges, he found that persons owning land to the north of the railway embankment agreed that the effect of the embankment had been beneficial to them. They were opposed to the idea of any more openings being made in it, on the ground that the benefit derived from its keeping out the Ganges flood more than compensated for its heading up floods from the rivers to the north. As regards the Gandak flood, Mr. Finnimore pointed out that the general fall of the country is from the Ganges to the Gandak, and that, when the latter is in high flood, the railway embankment in some places stops its spill, which otherwise would spread to the south when the Ganges was not in equally high flood. As, however, the Ganges flood always rises in August, and is during that month higher than the Gandak flood, he held that little, if any, benefit would accrue to the country north of the railway by providing sluices in the railway embankment to let the Gandak flood through, because they would have to be closed against the Ganges flood before the *bhadoi* crops were harvested. He further pointed out that from Lakshminia to Mansi the railway bank forms the only protection for the country to the north from the Ganges floods.

On receipt of his report, Government decided not to take further action, holding that an opening in the railway would have no appreciable effect on the level of the water to the south, as its level depends on the height of the Ganges flood. Speaking generally, there appears to be little doubt that, as the railway embankment is unbridged from a point well outside the district on the west to within a short distance from Khagaria (where there is a bridge over the Burh Gandak), and again from this point till it reaches the Kosi in the Purnea district, it serves to prevent the Ganges floods from spreading over a large area to the north, which was periodically flooded before the line was constructed.

CHAPTER VII.

RENTS, WAGES AND PRICES.

RENTS. Cash rents.

WHEN the original edition of this Gazetteer was published, a record-of rights had been prepared for the area north of the Ganges, and accurate statistics were available for this area; but in south Monghyr survey and settlement proceedings had been completed only in Government estates. The record-of-rights of the whole district has now been completed; and it is possible to give more accurate details of the general incidence of rent than could be given seventeen years ago. The survey and settlement proceedings in north Monghyr showed the following incidence of rents by the acre :—raiyats at fixed rates, Re. 1-9-8; settled and occupancy raiyats, Rs. 2-15-4 ; non-occupancy raiyats, Rs. 2-10-10. The average is Rs. 2-15-0 per acre, while under-raiyats pay Rs. 3-15-4 per acre. The incidence of rent rates paid in the different thanas by settled and occupancy raiyats, who form the bulk of the peasantry, is Rs. 3-11-1 per acre in Teghra, Rs. 3-6-9 in Begusarai and Rs. 2-6-9 in Gogri, the difference between the figures being largely accounted for by the extent to which the three thanas are liable to inundation. On the whole, rents are lower than in other districts of North Bihar, but this is due to the strikingly low rate of Gogri, in a large part of which there is still keen competition for tenants; those of thanas Teghra and Begusarai are more nearly on a level with those of adjoining districts. They are lenient in land held under indigo factories, and below the general average in Government estates.

In south Monghyr the incidence is as follows :—raiyats at fixed rates, Rs. 2-11-10; settled and occupancy raiyats, Rs. 3-8-5; non-occupancy raiyats, Rs. 3-5-4. Here, as in north Monghyr, settled and occupancy raiyats form the bulk of the peasantry; and the incidence of rent paid by them in the different thanas is as follows :—

	Rs.	a.	p.
Monghyr	4	0	1
Jamalpur	7	7	11
Kharagpur	5	4	1
Surajgarha	3	5	6
Lakhisarai	5	2	9
Sheikhpura	6	7	7
Sikandra	4	5	3
Jamui	2	15	10
Chakai	1	9	6

Produce-rents.

In north Monghyr the area for which rent is paid in kind is 6 per cent. of the land held by occupancy raiyats, and 7 per cent. of that held by non-occupancy raiyats. The actual area paying produce-rents held by these two classes of tenants is 35,894 acres and 2,453 acres, respectively. Cash rents are the rule in this area; but of late years landlords have been in the habit of making settlements on produce-rent of their *kamat* lands, including raiyats' holdings purchased in execution, because there is a prevalent idea that occupancy rights do not accrue in land held on produce-rent.

These conditions obtain also in those parts of south Monghyr in which cash-rents prevail; but in Kharagpur and west of the Kiul river there is a large area in which produce-rents prevail. The parganas west of the Kiul river formed part of the old district of Bihar; and we find here the same conditions as in the rest of *zila* Bihar. The following statement shows the percentage of area held on produce-rent in the different thanas :—

Thana.	Total area held by raiyats with rights of occupancy.	Area held on produce rent.	Percentage of area held on produce rent.	Percentage of net cropped area irrigated.
	Acres.	Acres.		
Monghyr ...	54,148	5,108	9	14
Jamalpur ...	853	91	11	37
Kharagpur ...	92,497	29,645	32	68
Surajgarha ...	48,288	8,877	18	22
Lakhisarai ...	69,668	21,815	31	30
Sheikhpura ...	104,623	65,240	62	60
Sikandra ...	110,498	82,505	74	70
Jamui ...	125,294	27,781	22	45
Chakai ...	153,849	2,718	2	11
SOUTH MONGHYR ...	759,718	243,780	32	42

SYSTEMS OF PRODUCE-RENTS. The principal systems of produce-rent found in the district are those in which the rent varies with the outturn, the proportion of the crop payable as rent being fixed; these are *danabandi* or appraisement, and *batai* or division of crops. There are also systems less commonly found, under which a fixed amount of grain is annually payable as rent, known as *mankhap* or *chauraha*.

***Hastobudi* crops.** When sugarcane is grown on land which is held under the *danabandi* or *batai* system, cash-rent is always paid for it, at certain customary rates which vary in different villages. Cash-rent was always paid when poppy was cultivated; but this crop is not now grown in Monghyr. Among these special crops we nearly always find potatoes; and in some cases rice is the only crop which is appraised or divided, all other crops being paid for at cash rates. Baden Powell in his *Land Systems of British India* speaks of these as *zabti* crops, interpreting the word as meaning sequestered or set aside. They are commonly known in Monghyr as *hastobudi* crops.

BATAI. The system which is most common in the district is that of *batai*, by which the crops are divided on the threshing-floor. The watchman, the reapers, and the weighman are paid from the whole crop, and the balance is divided in the appointed shares between landlord and tenant. Straw and husks, as under the appraisement system, are the property of the tenant. Raiyats prefer this to the appraisement system, partly because it affords opportunities for pilfering before division of the crop, but chiefly because they are less harassed under it than under the other system.

DANABANDI. The appraisement system (*danabandi*) is nearly as common as *batai*. Appraisement is made by an amin or *salis* in the presence of the raiyat, the *gumashta* and the *patwari*; there may be present also a *jaribkash* (usually the *gorait*), the *barahil* and the *jeth* raiyat or *mahatwara*. The appraisement is made by the amin, whose estimate is discussed by the assembled company. If the raiyat cannot be persuaded to accept the appraisement, the crop may be tested by cutting two *dhurs* (less often two *kattas*), one selected by the raiyat and the other by the landlord's party.

***Mafi* or *minhai*.** After appraisement, before the calculation of shares is made, a deduction of one or two seers in the maund is usually allowed to the raiyat. Buchanan mentions the allowance as covering the expense of harvest; it is possible that it was originally intended to cover the cost of cultivation, and that the allowance of two seers common in Monghyr district has been whittled down from an amount which was originally larger. This allowance is described as *mafi*, *minhai*, *tasalli*, or *garkti*. This last term is also applied to a special variation, where the remission is made by reducing the area of each field from the measured area by one *katha* in the *bigha*, which is equivalent to a twentieth of the measured area. This has the same effect as reducing the amount of crop appraised by two seers in the maund.

Cost of reaping, etc.

There is much variation in practice regarding payment of the cost of reaping and watching the crops. Sometimes the watchman is paid by both parties, sometimes by the landlord alone. The remuneration of the reaper is ordinarily one bundle out of every twenty-one, this is sometimes borne by both parties, sometimes by the raiyat alone.

The landlord's share.

Under the farming system which prevailed throughout the eighteenth century, though the raiyat's share after the customary deductions was nominally half, the farmers exacted *abwabs*, until as Ghulam Husain Khan said, the raiyats did not receive even a fourth of the crop.* In 1770 a large body of raiyats complained to the Revenue Council, with the result that on November the 5th of that year, it was determined that a clause should be inserted in the renters' covenants, by which they undertook, when collecting rent in kind, to leave 17½ seers in the maund to the raiyat.† This condition was retained in the covenants until 1788,‡ and we have here the evident origin of the tradition that nine-sixteenths of the crop is the normal rent in *zila* Bihar.

Abwabs.

The prevalent idea that the landlord's share in south Bihar is nine-sixteenths was not confirmed by the experience of the settlement officers who had to ascertain the actual realisations, since it was found in comparatively few instances that this proportion was realised. In very many cases it was found that the actual rent (the *asl jama* of the landlords' papers) was a half after the customary remissions, increased

* *See* Sir John Shore's Minute of April 2nd, 1788, Appendix XVII. Harington's Analysis, Vol. III.

† Harington's Analysis, Vol. II, page 246.

‡ Selections from the correspondence of the Revenue Chief, pages 11 and 161.

by various *abwabs*, of which the commonest was *dahiak*, most commonly of $1\frac{1}{4}$ seers or $2\frac{1}{2}$ seers in the maund of total produce after remission of *mafi*. Other *abwabs* were *nocha* (what is snatched away), *mangan* (a kind of forced benevolence), *amin. kharcha*, *kiali*, or *sonari* (payment for the weighmen) and *manseri*, one seer in the maund of the landlord's original demand (*asl jama*). As Mr. Hubback remarks, *kiali* is an unfair demand where the system is *danabandi*.*

Class of labourer.	1894-95.	1904-05.	1907-08	1924-25.
	A. P.	A. P.	A. P.	Rs. A. P.
Superior mason ...	4 0 to 5 0	6 0	6 0	1 14 0 to 3 5 4
Common „ ...	2 0 to 3 0	4 0	4 0	0 5 4 to 1 11 3
Superior carpenter	5 0	5 0 to 8 0	6 0 to 8 0	1 14 0 to 6 11 0
Common „ ...	3 0 to 4 0	4 0 to 5 0	5 0 to 6 0	0 5 3 to 1 11 0
Superior blacksmith	8 0	8 0	8 0	1 14 0
Common „ ...	4 0	6 0	6 0	1 0 0
Male cooly ...	2 0	2 0 to 3 0	3 0 to 4 0	0 8 0 to 0 12 0
Female „ ...	1 0	2 0	2 0 to 2 6	0 4 0 to 0 6 0
Boy „ ...	1 0	1 6	1 6 to 2 0	0 2 0 to 0 3 0

WAGES. The marginal table giving the daily wages paid for certain classes of labour sufficiently shows the upward tendency of wages in recent years. Agricultural labour is, however, generally paid in kind, the value of the grain thus given for a day's labour being from five to six annas.

Supply of labour. The following remarks regarding the supply of labour in Monghyr are quoted from Mr. Foley's *Report on Labour in Bengal (1906)*: "Emigrants to non-contiguous districts of Bengal in 1901 numbered 66,837, of whom 9,000 were in Calcutta, 4,000 in Burdwan, 3,000 in the 24-Parganas, 2,000 in Howrah and 1,000 in Hooghly. On the other hand, there were 6,000 in Dacca, 3,000 in Rangpur and 3,000 in Dinajpur, which shows that the emigration to Eastern Bengal was considerable. The Begusarai subdivision on the north of the river contains an extremely dense population, but the land here is very fertile. In this part the chief crop is the *rabi*, the *bhadoi* coming next, and the winter rice being comparatively

* This account of produce-rents in south Monghyr has been mainly compiled from the note of Mr. J. A. Hubback, I.C.S., which appears as Appendix W in the Final Report of the Survey and Settlement operations in south Monghyr.

unimportant. In the rest of the district the chief crop is the winter rice. A cooly earns from 2½ to 3 annas a day. Emigration from north of the river appears to be chiefly to Eastern Bengal, the people going away in November and December and returning after March. A good many from this part of the district appear also to be recruited by the jute presses in Eastern Bengal. The amount of labour obtainable seems to depend on the state of the crops, and varies from year to year; the number of landless labourers appears to be great, since labour is obtained from Begusarai at a cheap rate for Monghyr town; but in a good year the people are fairly well off, and might not care to migrate.

"With reference to the rest of the district, the northern part of the Sadr subdivision is not to be recommended as a recruiting ground, since the available labour is required in the district. The Jamalpur workshops run daily workmen's trains from Jamalpur, west to Kajra, north to Monghyr and east to Bariarpur, and are contemplating extending these. Messrs. Ambler and Company at Dharahra also need all the labour they can obtain and complain of its scarcity. In the west and south of the district, however, in thanas Sheikhpura, Sikandra, Jamui and Chakai, recruitment is to be encouraged, as the land is mostly poor; there are a large number of landless labourers, and the people generally are very badly off. There is some *rabi* in Sheikhpura, but the land is mostly under winter rice. Emigrants go freely to the mills from here and also (especially from the Chakai thana) to the coalfields of Giridih and Jherria, but probably considerably more labour could be obtained, especially for the mills and for handling goods if it were sent for. The castes which go to coal seem to be mostly Musahars, Dosadhs and Nunias, besides some Dharhis. Labour would probably be easily obtainable from January till the rains broke. This part of the Monghyr district, I think therefore, is to be recommended for labour of all kinds."

Kamiyas.

One class of labourers calls for special mention, the *kamiyas*, landless labourers who bind themselves to life-long service, farm servants given a small pittance but allowed a house. The following account given by the Collector, Mr. Lockwood, fifty years ago, may be quoted here. "The lands in this district are chiefly cultivated by *kamiyas*, who are in point of fact bondmen to the landholder. They belong to the lower castes, particularly Musahars and Dosadhs. I doubt whether there is a single Musahar in the district who is not a bondman. I have asked many of them and received the same reply from all. It appears to be the custom for every man, directly he arrives at the age of puberty and wants a wife, to receive a few rupees under the name of a loan from the village landholder, and execute a bond similar to that annexed, which I procured at a village near Jamui, and is a fair specimen of its class. It is said that not only are the bonds never liquidated, but that the landholder would refuse to accept the money if tendered; and, so far as I can learn, the simple clown gets so confused regarding what is due in shape of compound interest that it never enters his head to liquidate his debt. These bondmen are paid about Re. 1-6 per mensem, and the interest of the bond is also nominally taken out in work. The *kamiyas* are a lean race, and the wonder is how it is they manage to subsist with their scanty clothing and insufficient food. But their women all take their share in labour, and the Government roads, zamindari embankments, and wood-cutting keep them from starvation. I have on many occasions urged them to emigrate, but they plead that their masters will not let them go, and that they prefer the ills they have rather than to fly to others they know not of."

The bond referred to is quoted as a curiosity among contracts. It runs :—" Agreement between A. B., a Musahar of village Mablepur, *pargana* Parbatpur, *zila* Monghyr, on the one hand and Babu C. D., Rajput, on the other. In consideration of receiving Rs. 5 in cash, to celebrate his marriage, A. B., hereby binds himself to plough, sow, irrigate and reap the fields of C. D., and perform faithfully all the duties of a *kamiya* or bondman. The said A. B. binds himself to continue in the service of his master C. D., and never to refuse doing any work imposed on him. Morning and evening, day and night, he will be present and ready to work, and he will never absent himself even for a visit to a friend or relation without leave. If, on any occasion, the said A. B., should absent himself, that day's work will be placed to his debt, and he will be liable for such damages as Her Majesty's courts of law may direct. In addition to the above duties, the said A. B. binds himself to furnish the said C. D. with the following commodities, as may be directed by a council of peers of the said C. D.—thatching grass, bamboos, strings, wood and other things. This deed is executed in good faith, 22nd Asarh, 1265."

The agreements were such that they could not legally be enforced; but *kamiyas* did not know this, and they practically differed only from slaves in the fact that if they could obtain the amount of money advanced under the bond they could redeem themselves. The *Kamiauti* Agreements Act (VIII of 1920), makes such contracts void if the term for which labour is to be rendered exceeds one year, if the debt is not to be extinguished with the term, or if fair remuneration is not to be provided for the labour. The Act does not apply to agreements entered into by "skilled workmen" so that the old *kamiauti* conditions may still apply to labour rendered by such persons as Chamars.

PRICES.

The following table shows the average prices of staple food crops, published by authority of the Local Government under section 39 of the Bengal Tenancy Act, since the year 1895 :—

YEAR.	Monghyr.				Begusarai.				Jamui.			
	Wheat.		Rice.		Wheat.		Rice.		Wheat.		Rice.	
	S.	Ch.	S.	Ch.	S.	Ch.	S.	Ch.	S.	Ch.	S.	Ch.
1895	15	11¾	15	1¾	16	1	17	4	16	7½	16	11½
1896	12	2	11	15½	11	14¼	13	0⅓	11	13⅛	13	3⅔
1897	9	5	8	12½	8	14	8	12	9	0⅛	9	8½
1898	15	6⅔	12	14	14	14¾	13	12½	15	0	15	4
1899	16	11½	13	0⅔	17	11	14	4	17	6⅔	15	12
1900	13	8¼	11	11⅔	14	3	13	1⅔	13	10⅔	13	2¾
1901	13	7½	12	4¼	12	9	12	7	11	8	11	5⅓
1902	14	10	10	15	13	14½	12	4⅔	12	10	11	14⅔
1903	15	6½	12	3	16	3	15	1½	13	14⅔	13	7½
1904	17	7	12	14	17	15¼	14	11	17	5⅛	13	2
1905	15	0½	12	9	14	7	13	12	14	2¾	13	2¾
1906	11	7¾	8	14¾	10	5¼	9	0	11	6¾	9	12
1907	10	4¼	8	1¼	9	6¾	7	13¾	9	15	8	8¼
1908	8	9¾	7	5¼	8	3¼	7	14½	7	11	7	1¼
1909	9	15	9	9	9	4	8	15	9	8	9	2
1910	12	12	13	4	11	7	13	1	12	13	13	2
1911	14	11	13	5	13	11	12	12	14	6	13	3
1912	12	14	10	14	11	15	10	12	12	[illegible]	10	13
1913	10	7	7	7	10	1	7	14	10	1	8	6
1914	9	8	7	9	8	10	7	3	9	1	7	10
1915	8	3	7	3	7	14	6	15	7	3	7	4
1916	9	11	8	4	9	10	8	3	8	11	8	[illegible]
1917	7	6	10	7	10	6	10	5	10	11	11	[illegible]
1918	8	0	9	3	8	2	9	3	8	0	9	0
1919	6	10	5	15	6	10	5	6	5	4	5	5
1920	6	13	5	9	6	7	5	6	6	4	5	14
1921	6	3	5	9	6	1	5	13	6	4	6	3
1922	6	11	6	1	6	5	6	14	5	15	6	2
1923	8	12	7	0	8	2	7	9	8	2	7	14

MATERIAL CONDITION OF THE PEOPLE.

A special inquiry into the condition of the agricultural classes was instituted during the survey and settlement of North Monghyr, the result of which is summarized as follows by the Settlement Officer :—" It may, I think, be assumed that the average cultivator's family in north Monghyr has a fair margin beyond the mere minimum subsistence allowance : and any one who knows the area will readily recognize that without such a margin the effects of the frequent and disastrous floods would be felt much more severely than they are, and the vast sums of money lavished, more especially in the two western thanas, on litigation would not be forthcoming. The same fact is borne out by the figures showing the extent of indebtedness ; barely 4 per cent. of the raiyati holdings, and only $2\frac{1}{2}$ per cent. of the area covered by them, are affected in whole or in part by sales or mortgages with possession. The total amount of recorded indebtedness is under 4 lakhs, which represent less than one-thirtieth of the net annual profits of cultivation, or barely half-an-anna in the rupee.

" The general position may be summed up thus :—The cultivator in Teghra and Begusarai is well off in spite of high rents and extreme pressure of population on the soil, because the soil is highly productive, and especially favourable to the growth of the more valuable autumn and spring crops ; in Gogri rents are low, but land is abundant and gives a fair return without any great expenditure of money in cultivation. In both areas the effects of bad seasons more or less counter-balance one another, and even floods, which sweep away or ruin hundreds of autumn crops, have their compensating advantage in an improved spring crop."

So far as statistics of sales and mortgages of raiyati holdings in South Monghyr are concerned the figures obtained as a result of the enquiries made during survey and settlement proceedings in that area are equally reassuring, since only 3·67 per cent. of the area held by settled and occupancy raiyats had been the subject of mortgage or sale.

Co-operative Credit Societies.

The co-operative movement has made considerable progress in Monghyr district. There are Central co-operative Banks at Monghyr, Begusarai. Sheikhpura, Jamui, Khagaria, and Haveli Kharagpur. There were, at the end of 1924, 386 agricultural and eleven non-agricultural co-operative societies; and there were also ten Guarantee Unions to which 86 societies were affiliated. The Registrar of Co-operative Societies reports that the movement has gained a firm footing in the district, and that it is doing much to improve the moral and material condition of the people; since, apart from the promotion of financial independence among the people, the Banks and societies do their best to introduce improvements in agricultural methods, to spread primary education, and to encourage village sanitation.

CHAPTER VIII.

OCCUPATIONS, MANUFACTURES AND TRADES.

OCCUPATIONS.

ACCORDING to the statistics obtained at the census of 1921, 75 per cent. of the population are supported by agriculture and pastoral pursuits, 6·1 per cent. by industries, 5·9 per cent. by trade, and 1·4 per cent. by the professions. Of the agricultural population, 48 per cent. were returned as "actual workers", including 4,200 rent-receivers, 479,000 rent-payers, and 233,000 labourers. Of the industrial population 52 per cent. are classed as actual workers, including 8,000 potters, 4,200 cotton weavers, 4,100 carpenters and 3,000 shoemakers. Of the professional classes 45 per cent. are actual workers, including 4,300 priests, 800 teachers and 450 musicians. The trading classes including 10,000 milk vendors, 9,400 sellers of fruit and vegetables, 5,700 grocers, 5,000 grain dealers and 3,400 fish dealers. Among those engaged in other occupations are 64,000 general labourers. An indication of changing conditions may be seen in the fact that the 5,000 *palki*-bearers mentioned in the Gazetteer of 1909 had fallen to 687 in 1921.

Agricultural classes.

During the rule of the Mughals the largest landholders appear to have been Kazis, Malliks, Mirs, Mirzas and Shahs, to whom lands were given for services performed, whether military or civil. The south of the district was held mainly by the houses of Kharagpur and Gidhaur, which have played an important part in its history since early Muhammadan times and have done much to preserve the old order of things. The house of Kharagpur, however, has now completely declined, the last of the estates having been sold up by the Government and bought by the Maharaja of Darbhanga; and at present the principal landholders south of the Ganges are the Maharaja of Gidhaur and the purchasers of the Khaira estate. In course of time, owing to the prosperity of the professional and commercial classes, the operation of the sale laws, and the indebtedness of the landholding classes, other classes, such as Banwars, Agarwalas and Marwaris, have risen to the position of zamindars. There are also a few instances in which the mendicant or Sannyasi classes have acquired wealth and importance, as, for instance, the Mahants of Dularpur and of Suja. Their connection with the land is as a rule due orginally to the piety of their disciples, who endowed the *maths* of their spiritual *gurus* with extensive properties. Other zamindars of recent creation are indigo planters, who, beginning generally as tenure-holders, become in the end proprietors, and, following their calling, constitute a small class of cultivating zamindars.

Tenure-holders are principally Babhans, Rajputs, Kurmis, Sheikhs, Christian indigo planters and Ghatwals. The majority of them cultivate the best land in their tenure, while they settle the rest with others at advantageous rates of rent, so as to leave them a margin of profit after paying the zamindar his dues. The lessees (*thikadars*) were frequently occupancy raiyats in the beginning, who thought it best to invest their savings in a business in which they had special proficiency. It frequently happened also that the zamindar, being an absentee landholder, found that the most economical way of realizing rent from a distant zamindari was to give a lease of his land to persons who, being residents of the locality and cultivators themselves, had the confidence of the raiyats and could realize rent more easily than a *gumashta* or any other hireling deputed by the owner.

Among occupancy and non-occupancy raiyats Babhans, Goalas, Koiris, Dhanuks, Brahmans, Rajputs, Kurmis, Santals and the various classes of Muhammadans bulk largely. Agricultural labourers are mainly recruited from among the low castes, such as the Musahars, Bhuiyas and Dosadhs.

Industrial classes.

The principal castes engaged in producing fabrics and dress stuffs are the Jolahas and the weaver caste known as Tatwa or Tanti, a functional group developed under the pressure of the natural demand for woven cloth. The Dhunias card cotton; the Rangrez are dyers; and the Laheris make lac bangles; the Churihars also are bangle-makers; and the Patwa makes fancy silk strings and fringes. Numerous castes prepare food and drink. The Hindu Goalas and Muhammadan Ghosis are dealers in milk and curds; the Chiks and Kasais are butchers; the Telis are oil-pressers; the Kandus are grain-parchers; and the Halwais are confectioners. The Pasis tap palmyra and date-palm trees and sell toddy; the Sunris are manufacturers and vendors of country liquor; and the Kallas and Kalwars are Muhammadan distillers and liquor sellers. The Chamars are the village tanners; while the Dabgars make skin or leather vessels used for *ghi*, oil, country saddles and the like. The Barhis are carpenters; and several castes work in bamboo, such as the Turis, Karangas (or Bangars), and Doms. Other bamboo-workers are the Mahlis, a Dravidian caste of labourers and palanquin bearers, who came originally from Chota Nagpur. The Barais make the leaf platters used at Hindu ceremonies and festivals, and also make and carry torches. Among workers in minerals may be mentioned the Nunia or saltpetre-maker, Sonar or goldsmith, Lohar or blacksmith, Thatera and Kasera (also called Kansari and Kansabanik) or brazier, and Kumhar or potter.

Commercial classes. The chief commercial classes are Agarwalas, Agraharis, Marwaris and Baniyas. Among Agarwalas the largest proportion are engaged in banking, trade, petty money-lending and similar pursuits. A few are zamindars and holders of large tenures, but in most cases their connection with land may be traced to a profitable mortgage on the estate of a hereditary landholder. The poorer members of the caste find employment as brokers, servants, and workers in gold and silver embroidery, and they take to any respectable pursuits except agriculture. The Agraharis are another trading caste, among whom the women are not secluded as among the Agarwalas, but take part in the business of their husbands by selling rice, flour, etc. Their business is generally that of tradesmen rather than that of bankers. Most of the Marwaris, who are generally traders from Rajputana and Central India, are bankers and traders, and are usually Jainas by religion. Among petty traders may be mentioned the Rauniars and various Baniya groups dealing in cloth and grain, the Tambuli or seller of betel-leaf, and the Kunjra or green-grocer.

Professional classes. The Brahmans are naturally the most important of the classes engaged in a religious profession. They perform the religious ceremonies of the upper classes of Hindus as well as of those Brahmans who, being engaged in other pursuits, are unable to attend to their own spiritual needs without the aid of some one especially skilled in the Sastras. Among the literary classes the Kayasths rank first. Clerical work is believed to be the original and characteristic occupation of the caste, and an illiterate Kayasth is looked upon as a creature who does not justify his existence. Kayasth tradition, however, puts a very liberal construction on the expression clerical work, and includes in it not merely clerical pursuits of a subordinate character, but business management; and estate *tahsildars* and *patwaris* are almost invariably Kayasths. Some Kayasths are in possession of considerable zamindaris and tenures of substantial value, while comparatively few of them are to be found among the lower grade of cultivators. They form the majority of the school teachers; while the Sheikh, Saiyad, Mughal and Pathan are the classes from which the Muhammadan *maulvis* are principally drawn. The teaching of the latter consists mainly in the exposition of religious texts, and the language taught is Arabic, Persian or Urdu. Medical practitioners are recruited from Hindu Baidyas and Muhammadan Sheikhs and Saiyads, the medical science practised by the Sheikhs and the Saiyads being known as the *hakimi* system of treatment, while the Baidyas practise the Hindu system of medicine. The legal practitioners, i.e. pleaders and *mukhtars*, are mostly Bengalis and local Kayasths. The Kayasths form the majority of lower grade practitioners, and in the subtlety and skill which they frequently display seem to be no unequal match to the Bengalis, who form the majority of the higher grade practitioners of the district. The pugnacious tendencies of the people, and the pleasure which the landholding classes find in the excitement of litigation, give ample employment to legal practitioners; and the profession has come to be regarded as one of the easiest means of acquiring a good social position, a competence, and, with good fortune, wealth.

MANUFACTURES.

The skill of the artizans of Monghyr and the excellence of their work have long had more than a local reputation. One of the earliest products of the district now extant appears to be the black stone throne or *masnad* of the Nawab Nazims of Bengal, Bihar and Orissa, which bears on one of the 16 facets into which the rim is cut an inscription in Persian to the effect that " this auspicious throne was made at Monghyr in Bihar by the humblest of slaves, Khaja Nagar of Bokhara, in 1052 Hijra," i.e., 1643 A.D. Later, in the 18th century, when Mir Kasim Ali made Monghyr his capital and set up an arsenal there, the manufacture of iron-work appears to have flourished. Raymond (Mustapha Khan), the translator of the *Sair-ul-Mutakharin* (cir. 1786), indeed says:—" The European reader may possibly hear with surprise that the fire-locks manufactured at Monghyr proved better than the best Tower-proofs sent to India for the Company's use; and such was the opinion which the English officers gave when they made the comparison by order of the Council of Calcutta. The flints were all Rajmahal agates, and their metal more mellow."

Mr. Twining, who visited Monghyr in 1794, when on his way up the Ganges with the Commander-in-Chief, Sir Ralph Abercromby, found much to admire in the iron-ware of the Monghyr workmen. "Iron ore," he says, "is more precious than gold to the industrious inhabitants of Monghyr, sustaining, by its abundant and cheap supplies, an extensive manufactory of iron utensils of almost every description. A traveller from Europe is surprised to see presented to him for sale in the interior of India, knives and forks, saucepans, gridirons, and tea-kettles, which would not discredit an ironmonger's shop in London. But it is not the kitchen alone that Monghyr is able to supply, possessing an excellent fabrication of cabinet-ware, chairs, tables, sofas, cots, bedsteads, drawers, etc.—all made of a handsome black wood resembling ebony. The chairs and sofas, in particular, with cane bottoms, are extremely neat, and scarcely inferior to the European models after which they are made. The commerce of Monghyr in these articles is very extensive". He also informs us that the mineral water obtained from the hot springs of Sitakund was in great request, its principal virtue being "its purity and consequent wholesomeness, and also its property of remaining good for a great length of time in casks or bottles. For these reasons it is not only much drunk by the wealthier inhabitants of Monghyr, or such as can afford to send for it, but is often despatched to Calcutta for the use of persons about to undertake a long sea voyage. On these occasions it is either sent off in large jars, or in bottles filled at the spring. I was told that in the latter case it would preserve its tasteless purity for more than a twelve month. Recollecting, when I heard this, the offensive water of the ship on my way to India, I resolved, if it pleased God that I ever returned to England, I would take with me a few dozens of the water of Sitakund".*

From the journal of Bishop Heber, who visited Monghyr in 1824, we find that the industries of Monghyr were still flourishing. "I was surprised," he wrote, "at the neatness of the kettles, tea-trays, guns, pistols, toasting forks, cutlery and other things of the sort, which may be procured in this tiny Birmingham. The only thing which appears to be wanting to make their steel excellent is a better manner of

* T. Twining *Travels in India a hundred years ago*, 1893.

smelting and a more liberal use of charcoal and the hammer. As it is, their guns are very apt to burst and their knives to break, precisely the faults which from want of capital beset the work of inferior artists in England. The extent, however, to which these people carry on their manufactures and the closeness with which they imitate English patterns show plainly how popular those patterns are ". The mineral water of Sitakund was still exported, some persons in Calcutta drinking nothing else, while muskets and pistols were made and sold at cheap prices, besides spears, which were even cheaper, one of the best only costing Re. 1-4. The Bishop purchased a number of the latter for his servants, so that his cabin looked like " a museum of Eastern weapons ". His narrative also introduces us to another industry which has now fortunately disappeared, viz., the production of talc for use in windows, blocks of talc, which divided easily into thin but tough laminæ as transparent as isinglass, being obtained from the Kharagpur Hills. Thirty years before his visit this was the only approach to glass usually seen in windows of houses, even of those belonging to Europeans.*

Many of the industries mentioned by the early writers still survive and flourish; and to them have been added large and important industries, such as stone quarrying and the manufacture of locomotives at the great workshops of Jamalpur.

Railway workshops. The largest industrial concern in the district consists of the workshops of the East Indian Railway at Jamalpur. These workshops, which are among the largest in India, have been established for the construction and repair of rolling stock and plant connected with the railway, and all the constituent parts of a locomotive engine can be constructed there. The Iron Foundries are capable of a monthly outturn of over three thousand tons of castings, while the Steel Foundry can turn out eight hundred tons of steel, and the Rolling Mill a thousand tons of rolled iron and steel bars in a month. The works cover an area of 99 acres, of which 26 acres are roofed over, and they afford employment to 278 Europeans and 11,500 Indians. The value of the annual outturn is about 1½ crores of rupees.

Slate quarries. The following account of the slate quarries of Monghyr is given by Professor V. Ball in the *Economic Geology of India:* " In the transition rocks of the Kharagpur Hills there

---

* R. Heber, *Narrative of a Journey through the Upper Provinces of India,* 1827.

is a band of slates, from 6 to 12 feet thick, which is traceable for many miles on the northern margin of the hills between Rishikund on the east and the Gaura and Amrasani *kols* on the west. Being for the most part vertical, it can only be worked by steps on the side of the hills, or by actual mines. For many centuries this slate has been worked by the natives more after the fashion ordinarily practised with reference to veins of metallic ore than to deposit of slate; the petty mines and quarries serve to produce an amount of slate which is equal to the demand, but the cost is certainly high. Its chief employment is for roofing instead of tiles, and a large quantity is so used at Monghyr and the neighbouring town; curry platters, etc., are also made from this stone to some extent." This account was written many years ago before the quarries were developed.

There are now six slate quarries in the Khäragpur Hills, at Maruk, Sukhal, Tikai, Garia, Amrasani and Sitakobar. These quarries have been worked since 1864 by Messrs. Ambler and Company, who purchased the Basauni indigo factory from the late Mr. Deare and converted it into a slate factory. The stone quarried is a slightly metamorphosed phyllite, which gives excellent roofing slates, and also produces fine slabs which are extensively used for electrical purposes. The slabs after being sawn and polished are black enamelled, and are made up into switch-boards, fuse bases, and knife switches. In 1914 the Company began to manufacture school slates, which are prepared in a fully-equipped workshop where the slates are polished and fitted with wooden frames. The output fluctuates; the average outturn during the last five years was 2,554 tons of roofing slates, and 156,988 school slates. The labour force employed in 1924 was 300 below ground and 107 above ground. Wages range from 4 to 10 annas a day; but many labourers on piece-work earn from a rupee to Re. 1-8-0 a day. In 1913 the concern was formed into a limited liability company under the name of Ambler's Slate and Stone Company, Limited.

Mica mines. The south of the district contains part of the Bengal mica-producing area, which roughly coincides with a great belt of schists and associated gneissose granite, some 12 miles broad and about 60 miles long, which stretches from Bendi in Hazaribagh district through the south-eastern corner of the Gaya district north-eastwards to near Nawadih (Jhajha) on the East Indian Railway in Monghyr. Here there are six mines at work, at Burhia, Mahgain, Srikrishna, Ganda, Sahaja Pubari, and Bhukle in thana Jhajha, and at Bichway in thana Sikandra. The average yield during the five years ending in 1924-5 was about 9½ tons, as compared with nearly 14½ tons in the previous quinquennium, the decline being attributed chiefly to the product being outclassed by that of the Hazaribagh mines. In 1924 the outturn was 10 tons, and the total number of employés was 315. The men employed are local residents who come daily from their homes in the adjoining villages. Wages range from 5 to 6 annas daily for men (averaging 5½ annas), and are 3 to 4 annas for women and 2 annas a day for children. The largest mines are at Bhukle, where there are six excavations, varying in depth from 9 to 69 feet, and at Mahgain, where there are six excavations of 6 to 30 feet. The process of mining is of a primitive character. The mica is generally discovered at the exposed outcrop of a pegmatite vein on a hill face and is followed from "book" to "book" by underhand stoping, which results in tortuous worm-like holes. The whole material extracted, consisting of mica, rubbish and under-ground water, is brought to the surface by a string of coolies working hand over hand on rudely constructed bamboo ladders.

There are some other mica mines not being worked at present which belong to the Maharaja of Gidhaur. These are to be found in mauza Behara.

Other minerals.

Not far from the slate quarries of Dharahra is a hill of quartzose granite or millstone grit, the stone extracted from which is utilized for grindstones. Lime is also exported from Monghyr in considerable quantities. The coarsest kind is made from *kankar* or nodular limestone, which appears to have been washed from the limestone formations of the Himalayas and deposited all over Bihar during the oscillations of the rivers by which it is traversed. The stone is burnt in kilns in order to expel the carbonic acid gas, and the residue is collected and sold at a low price. The best lime, however, is made from the shells of fresh-water molluscs, which are found in great abundance in the marshes.

Indigo manufacture.

In Sir William Hunter's Statistical Account of Bengal (published in 1877) it is stated:—"The great European industry connected with the preparation of indigo has now for a long time taken the first place amongst manufactures in this district, and still holds that position. The area under indigo cultivation is estimated at 10,000 acres, and the outturn at 4,000 maunds, or about 143 tons ". Sir William Hunter then proceeds to quote a return submitted by the officer in charge of the Begusarai subdivision, " in which nearly all the chief factories are situated ", which shows that in 1869 five concerns were at work, viz., Manjhaul, Begusarai, Bhagwanpur, Begamsarai and Daulatpur. These concerns cultivated 19,500 *bighas*, employed 6,400 to 8,400 hands, and produced 1,590 maunds of indigo; but the outturn in that year was very much less than in average years. In an article on the " Distribution of Indigo in Bengal " published in the Statistical Reporter of 1877 we find a fuller account of the industry, which is as follows :—

" Indigo cultivation in Monghyr is believed to be decreasing, and is now almost entirely confined to the tract of country lying north of the Ganges. On the south of the river there are only four factories—two European and two native—and their manufacture for some time past has been very inconsiderable. North of the Ganges eleven factories are at work—eight being under European, and three under Indian management. The area cultivated with indigo in the whole of the district may be stated as 15,000 acres, producing on an average 1,500 maunds of manufactured dye. It is said that $5\frac{1}{2}$ lakhs of rupees are yearly expended on indigo cultivation in the district. In Monghyr, as elsewhere, indigo is grown on the best high lands that are available, and the cultivation is unpopular with the raiyats, as displacing their most profitable crops. The zamindars, however, are favourably inclined towards the planters, who pay a high rent for indigo lands; while the low, landless castes, such as Musahars, Doms and Dosadhs, get better wages for work on indigo cultivation and manufacture than they could hope for in other lines."

Even as late as 1901-2 there were the following factories and outworks : (1) Begamsarai, with outworks at Teghra, Barauni, Nawada and Jhamtia; (2) Bhagwanpur, having outworks at Agapur, Surajpura and Kirtaul; (3) Harrakh, with outworks at Masnadpur, Mahna, Dhabauli and Mati Hasir; (4) Manjhaul, with outworks at Bishunpur, Birpur and Garhpura; (5) Sisauni, with outworks at Bandwar, Gamharia and Kamalpur; (6) Nayagaon; and (7) Sadanandpur, with outworks at Hitanpur and Samastipur. There were also four outworks of factories in Darbhanga, viz., Meghaul and Ramnagar, outworks of Daulatpur; Malipur, an outwork of Mangalgarh; and Gobindpur, an outwork of Dalsinghsarai. Even in the short time which has since elapsed, the above list is mainly interesting from the point of view of a local historian. The number of ruined factory buildings which one sees in North Monghyr is a melancholy sight. The industry is declining owing to adverse seasons and the competition of the artificial product. Most of the planters have disappeared, and those left are taking up zamindari and the cultivation of country crops as well as indigo.

Subdivision.	Factory.	Outwork.
Begusarai ...	Daulatpur ..	...
	Manjhaul ...	Bandwar.
		Bishunpur.
		Gamharia.
		Garhpura.
		Sisauni.
Monghyr ...	Digaon ...	...
	Sangrampur	...

The marginal table shows the factories now working. The two factories in the Monghyr subdivision are both in thana Kharagpur.

A modern industry in Monghyr has been introduced by the Peninsular Tobacco Company, which was formed for the purpose of manufacturing cigarettes from Indian leaf tobacco. Monghyr was selected on account of its reputation as one of the healthiest places in Bihar, within easy reach of a considerable tobacco area, and having advantages of rail and river transport. The Factory is situated in the Basdeopur ward of Monghyr, covering with its outbuildings a site of about sixty acres. The main building is four hundred feet long by sixty broad, with a chimney 108 feet high which forms a prominent landmark; and a great number of godowns have been built for the storage of leaf tobacco, which are rendered necessary by the fact that leaf tobacco matures with age, and its smoking qualities are thereby improved. There is a very complete carpentering plant, equipped for converting rough logs of timber into finished packing cases. The whole of the machinery is driven by electrical power, for generating which

Tobacco manufacture.

there are five Lancashire boilers and three sets of high pressure compound engines, each of a capacity of 325-B.H.P. Water for the use of the whole factory, and for reserve against the danger of fire, is obtained from the Ganges, and stored in three reservoirs, each having a capacity of 100,000 gallons.

Residential buildings have been erected within and near to the factory compound for the European staff, who instruct the Indian labour in the work. The labour force consists of about three thousand people, who work under very good conditions in excellently lighted and ventilated premises. The factory, which is equipped with the most modern machinery, and managed very efficiently, has been working since May of 1908; experiments are made daily in the constant effort to improve the smoking qualities of its manufactures.

Iron work.

One of the oldest and most interesting industries of Monghyr is iron and steel work, in particular the manufacture of guns, which dates back to the time when the Nawab, Kasim Ali Khan, made the town his headquarters.

There is a general belief that this has long been a declining industry. Forty years ago, for instance, it was reported that it formerly flourished, in consequence of the iron produced by the smelters in the Kharagpur Hills, but languished when iron articles began to be imported, the iron workers finding more lucrative employment in the workshops of the East Indian Railway at Jamalpur. It is doubtful, however, whether the industry is quite so decadent as is supposed. At any rate, it appears to be more flourishing now than it was 50 years ago, for in 1876 an article on Monghyr as one of the trade centres of Bengal stated specifically—"Only four persons during the past year took out licenses for the manufacture of firearms, and the amount of guns and pistols turned out during that time did not, it is alleged, exceed one hundred."* From Mr. Collin's *Report on Arts and Industries in Bengal*, it is clear that by 1890 the industry had revived, the average number of guns made annually being over 2,000. "Formerly", Mr. Collin says, "there were only five shops which sold guns at the surrounding *melas*. This was stopped under the Arms Act, and the manufacture was almost extinguished. A demand for cheap guns, however, still continued, and was supplied from abroad though Calcutta. The Monghyr gunmakers

* Statistical Reporter, May, 1876.

continued to protest to the authorities against the disabilities imposed on their trade, and the export from Calcutta of foreign guns was, it is said, stopped. However this may have been, a demand sprang up for Monghyr guns, and now in the place of five shops there are twenty-five. They make single-barrelled guns at about Rs. 10 each. They no longer make their own iron, but import it. The barrels are made of iron rolled into a cylindrical shape, welded together and then bored. All the parts of the lock, including the small screws, are home-made. The old gunmakers object to the new shops which have sprung up and say that guns are now made so cheap that they cannot be safe. They even applied for an Inspector to test the barrels, lest some accidents from bursting barrels should discredit the whole trade ". This suggestion, it may be added, was not accepted by the then Commissioner of Bhagalpur on the ground that it would be a needless interference with the industry.

Coming to more recent times, the returns shew that in the five years ending in 1899-1900 the number of guns and pistols exported was 11,575, representing an outturn of 2,325 per annum, which is even better than the figure quoted for 1890. In the next quinquennium, however, i.e., in the five years ending in 1904-5, the number fell to 5,825, this marked decrease being attributed to the falling off in the demand, owing to the greater care taken in the issue of gun licenses, to the competition of superior weapons of European and American manufacture at cheaper rates, and to the high wages which the operatives are able to obtain at the railway workshops of Jamalpur. The present condition of the industry is described as follows by Mr. J. G. Cumming, in his *Review of the Industrial Position and Prospects in Bengal in* 1908: " The industry still thrives in Cossimbazar, so called after the Nawab who brought a carpenter from Delhi for making gun stocks, from which small beginning this gun industry arose. There used to be 22 shops, but there are now only 13. The annual outturn of guns and pistols rose from 2,000 in 1890 to 3,000 in 1897. At present, the number manufactured annually is about 800. It is really wonderful what fine results can be obtained by means of the crudest of methods and the simplest of appliances. The gunsmiths complained of loss of custom; but they appear to be sweated by the wholesale Indian firms in Calcutta, about six in number, who buy their guns at Rs. 8 to Rs. 10 for a single-barrel and Rs. 15 to Rs. 20 for a double-barrel gun. The guns made to order fetch higher values according to their quality. A single-barrel breech-loader costs locally Rs. 30, a double-barrel breech-loader Rs. 50. A good workman can earn as much as Rs. 30 a month." Not only guns, but also pistols and sword-sticks are turned out by the artisans.

The following account of the method of manufacture is quoted from *A Monograph on Iron and Steel Work in the Province of Bengal* (Calcutta, 1907), by Mr. E. R. Watson :—
"Gun-barrels of three kinds are made, viz., plain, marked with simple twist (*mowa*), and damascened. The simple twist is a more or less regular spiral mark running round the barrel, the marking being in the metal just as in damascened work. The damascened barrel is marked all over with small spirals of about $\frac{1}{4}$ inch diameter. To make a plain barrel, a piece of Swedish iron bar is taken and hammered into a strip about 6 feet long, 1 inch wide and $\frac{1}{4}$ to $\frac{1}{2}$ inch thick. This is then hammered into a close spiral, such as would be formed by winding the strip round a straight rod. Neighbouring coils of the spiral are touching. Thus, a rough tube is made, the bore being considerably less than required in the finished barrel. By heating and hammering, the coils of the spiral are welded together, and the wall of the tube has now become solid. To prevent the iron being spoilt by so many heatings, it is generally covered with mud before being put in the fire. In forging the ends of the tube a mandril is inserted into the bore to prevent the lumen closing up. The tube is now bored, and for this purpose an implement is used which may be likened to a large railway carriage key. The barrel is fixed firmly in position, passing through a hole in a large post, which is itself firmly fixed in the ground. A man now inserts a borer of small bore into the barrel and gradually bores through the barrel. This operation scrapes the sides of the lumen and makes the bore slightly greater and more uniform. A slightly larger borer is now inserted and the operation repeated, and gradually the bore is made larger and more uniform, until the size is attained. This operation must be done gradually and generally takes a man three days. The outside of the barrel is now filed up to the desired shape.

"To make a barrel with the simple twist marking, a number of strips of Swedish iron—say, about $\frac{1}{2}$ to 1 inch wide,

¼ inch thick and 8 to 10 inches long—are laid alternating with the same number of soft steel strips of the same width and length, the steel used for this purpose being the bands taken from bales of imported cotton goods, etc. About 16 of these strips are piled together and held together by a soft iron strip, which is welded round them. We thus have a bundle about 8 or 10 inches long, 3 inches wide and ½ inch to 1 inch deep, showing the edges of the 16 strips at what may be called the surfaces of the bundles. This is now heated and gradually hammered into a strip, about 6 feet long, 1 inch wide and ¼ to ½ inch thick, in such a way that the lines along which the alternate layers of iron and steel have welded run the length of the strip. This strip is now welded with one of soft iron of similar dimensions, and the strip thus obtained is used for making a barrel in the same way as described already for the preparation of a plain barrel. Of course, the composite layer is kept outermost.

"To make a damascened barrel a number (say, eight) of composite strips are prepared in the manner already described, but they are made of smaller size. Each strip is then twisted many times until it looks like a long screw—say, 3 feet long, ½ inch diameter and with a ½ inch pitch. These eight screw-like rods are tied together and forged out into a long strip about 1 inch wide and ¼ inch thick. This is welded to a soft iron strip of similar dimensions, and the composite strip is made into a barrel in the way already described. Of course here also the composite layer is kept outermost. After the barrel has been filed up true and polished on the outside, the markings are brought out by the application of a solution of chemicals known as 'English mixture'

"The guns usually produced are single-barrel muzzle-loading 12-bore shot guns fired by a cap, the cap-nipple being at the side of the breech-piece. The breech-piece is forged and filed out of one piece of soft iron, and is quite a complicated piece of work. This is screwed on to the barrel, the screw thread being made by English taps and dies. The largest shop in Monghyr belong to one Burri Mistri, who is somewhat more advanced than the rest of the gunmakers. He can make a very good imitation of almost any gun you will give him—double-barrelled breech-loaders with choke-bores, etc. I noticed in his shop a tool for finally polishing the interior of the barrel, which was very similar to the tool used at the Government Small Arms Factory at Ishapore for the same purpose, viz., a hard steel tool with a rectangular polished edge, which is packed with pieces of horn, paper, etc., to fit the bore. He can "blue" plain barrels, and temper the lock, triggers, etc., so that they show a play of colours. The locks are generally chased with ornamentation after English patterns."

Iron smelting.

Iron smelting was formerly carried on fairly extensively in the Kharagpur Hills, but it was given up there in consequence of the import of foreign iron, the smelters finding employment, and getting better wages, in the railway works at Jamalpur. It is still carried on, however, on a very small scale in the extreme south in the Chakai thana. The smelters, who are called Kol Majhis, obtain iron ore from the beds of the hill streams and smelt it in rude furnaces in a primitive manner. The industry is almost dead, but the Santal still prefers the iron thus prepared for the head of his arrow. The number of iron smelters returned at the census of 1901 was 640.

Gold and silver work.

Gold and silver work is carried on by the local Sonars on a small scale, but their work is of a very ordinary character. More finished articles are turned out at Kharagpur, where the artisans are said to have been introduced by the old Rajas of Kharagpur. The following account of the industry at this place is quoted from *A Monograph on Gold and Silver Work in the Bengal Presidency* (Calcutta, 1905), by Mr. D. N. Mookerji. "Of the 200 souls now in the town, about one-sixth are actively engaged in the trade, out of which they make only a moderate living. That their profit is small may be concluded from the migration of several of the younger men to larger towns where their skill is better appreciated and more liberally recompensed. The majority of the jewels now made are bracelets, anklets with innumerable pear-shaped bells that tinkle musically, necklaces, earrings, belts and ornaments for the hair and forehead. A favourite means of enhancing the effect of the ornaments is to insert small cubes of gold and silver in the design. The surfaces of these cubes are brightly polished to resemble precious stones, and have an altogether dazzling effect when seen in the mass. Another equally common trick is to intersperse the jewel with small pellets of burnished silver. *Atardans* are also made,

"But the thing for which Kharagpur is most famous is the gold and silver fish with a small cavity between the head and the body * used for storing perfumes. Large specimens of these fish are also occasionally used as caskets to present illuminated addresses in. The chief peculiarities of the fish are their extreme lightness and flexibility. The body consists of thinly beaten out ovals overlapping each other. One edge of the oval is scalloped to resemble the scales of a fish, the plain half being hidden by the scalloped edge of the succeeding ellipse. These are loosely fitted on to each other and are kept together by fine wires running from the neck to the tail of the fish. As alternate scales are gold-washed, the combined effect of the gold and silver scales flashing in the sunlight is very pretty. The fish is usually built up from the tail, each successive ring being bigger than the last, till the required length is attained, and the head is then fastened to the last ring by two small hinges."

Wood work.

The carpenters and cabinet-makers of Monghyr have great skill in making inlaid ebony cases, necklaces of betelnut wood, palm wood, horn and ebony, brooches and bracelets of antiquated shape and pattern. Besides these, they carve articles of furniture, and make boxes, walking sticks, and other small articles, which are inlaid with patterns in horn and ivory. But the Monghyr carpenters not only excel in making inland cabinet-ware: they are equally clever in turning out chairs, tables, and almirahs. The kinds of wood chiefly used by the Monghyr carpenters are ebony, *paisar* (*Pterocarpus marsupium*), *sisu* (*Dalbergia sissoo*) and *kathal* (*Artocarpus integrifolia*).

The following account of the industry is quoted from *A Monograph of Wood Carving in Bengal* (Calcutta, 1903) by Chevalier O. Ghilardi: "I have observed that the wood-carving used for the frontages of houses and for their interior decoration is of very poor quality indeed. The reason is easily found in the fact that wood-carving, as it is understood in Bankipore, Patna and Gaya, is not carried on here. In reality, there is no wood-carving in Monghyr but only inlaid work, for which there is a fair demand and clever artisans. The small quantity of carving necessary in this inlaid work on small articles of daily use is made on ebony or mahogany

* The District Magistrate reports that this industry, and that of the cabinet-makers in Monghyr, are practically defunct.

wood, the depth of the carving not exceeding the thickness of a rupee; and therefore the work of the local artisans is not employed on heavy carving in *paisar* or *sisu* wood, such as pillars, architraves, friezes, doors, etc. The Monghyr carving has a certain reputation for the embellishment of inlaid work, and the craft has been handed down from father to son for several generations. The few families practising it were established here in the middle of the 18th century, and they were brought to Monghyr, as the tradition goes, by Mir Kasim with his gunmakers. It may be supposed then that the wood fittings of the guns were the only articles worked upon in the very beginning of the industry by these *mistries* who, it is known, excelled in making rich inlaid work on the gun carriages. When the demand for such work ceased in the course of time, the artisans devoted themselves to other branches of the industry, and applied their inlaid work to objects such as those in use at the present day."

Chevalier Ghilardi mentions the work of two men who were considered "the local celebrities of the profession" in Monghyr, and were specially selected to work on the exhibits of the Darbar Exhibition at Delhi. "These men doubtless possess talent, but they lack originality. The forms of the articles they make are in each case always the same, and the decorative motives are never changed. Whether it be the cabinet, the table, boxes, brackets, ink-stands, stools, frames, etc., the pieces have always the same form, the sole difference being confined to the proportions. When we have seen the small stock of these goods usually kept in a shop, all the other collections are mere repetitions, with more or less finish, which latter factor influences the cost of production."

Other industries.

Among other minor industries may be mentioned basket-making, which is mostly carried on by the low caste of Doms, who make neat baskets from wheat straw and fibres of various grasses. Fancy baskets of a better class are also made of *siki* grass and fine bamboo chips woven together with silk or cotton thread; this basket-ware attracted attention at the Melbourne Exhibition of 1881. Pottery-making is a regular village industry, the articles produced being mostly porous water vessels. Cotton weaving is also carried on in the villages, coarse cloth being produced, but the industry is gradually dying out owing to the competition of machine-made piece-goods. Coarse blankets are woven by a few families of Gareris, and there are a few dyers and calico-branders, the centre of the latter industry being at Sheikhpura. The same place is also noted for the manufacture of tubes (*naicha*) for the hookah or Indian pipe. Saltpetre is manufactured on a very small scale, and the industry is not flourishing. There is a small manufacture of molasses, the *gur* of Gidhaur having a special reputation and selling generally at higher prices than that manufactured in other places. Aerated water is made from the mineral springs near Monghyr and exported.

TRADE.

The district is favourably situated for trade both by rail and river. The most important river marts are Monghyr, Simaria, and Gogri on the Ganges, and Khagaria on the Gandak. Monghyr, Barhiya, Lakhisarai, Jamalpur, Sheikhpura and Bariarpur are the chief centres of trade on the railway, while a considerable volume of traffic passes *via* Tarapur to Sultanganj station in the Bhagalpur district. Monghyr being almost entirely an agricultural district, its export trade consists mainly of agricultural products, the chief of which are various food-grains, oilseeds, gram and pulses, wheat and mustard, chillies and tobacco leaf. There is also a considerable export of raw sugar and of *ghi;* and a small trade is carried on in lime, saltpetre and hides, the trade in hides being comparatively recent. Slate is exported from the quarries in the Kharagpur Hills, and much of the *khas-khas* grass used for *tattis* elsewhere is exported from Monghyr, being collected in the low-lying Pharkiya *pargana* by the Khanjars or string-makers. Thatching grass is also collected and exported from the same locality. The principal imports are salt, piece-goods and coal, the greater part of the coal being sent to the railway workshops at Jamalpur. Other imports are rice, refined sugar, cotton twist and yarn, and kerosene oil.

From the south of the district *mahua* is exported in large quantities for use in distilleries. *Sabe* grass is cut by Santals and Naiyas in the Kharagpur and Mahesri Hills and is sent to the mills for paper manufacture. Bamboos are cut in the Batia and Mahesri Hills during January and March, and make strong *lathis*. They are sold for half an anna or one anna each to pilgrims, who throng the road at this season on their way to Baidyanath. Each pilgrim generally takes back with him as many as he can conveniently carry, and probably disposes of them for a fair profit to his up-country brethren. Mica is mined in Mahesri and exported *via* Tisri and Giridih. At Nawadih railway station (Jhajha) there is a thriving trade in *biris*, i.e., cigarettes made of a little tobacco rolled up in the leaf of the *kend* tree, which is supposed to be very aromatic when used in this way. Santals and Naiyas get one pice per *binda* or small bundle of these leaves, and they can pluck and bring in six to eight *bindas* a day. The tobacco is imported from Calcutta and from the north of the district. The actual makers of the cigarettes get 4 annas per thousand, which they can earn in a day, and the manufacturer gets Re. 1 to Re. 1-4-0 per thousand for the cigarettes from Calcutta firms. The leaves of the *sal* tree are also exported in large quantities for the preparation of leaf-plates, which refreshment vendors at railway stations use for serving out fried gram and other delicacies to hungry travellers.

CHAPTER IX.

MEANS OF COMMUNICATION.

Development of communications.

Till within comparatively recent times the chief highways of commerce, and the main routes by which this district was connected with other parts of India, were the river Ganges and the road running along its southern bank. Local tradition ascribes the latter to the Emperor Sher Shah, and it is known to have been the highway along which Muhammadan armies passed in their marches to and from Bengal and North-Western India. The Ganges, however, appears to have been used far more by travellers, several of whom have left accounts of the journey up and down the river. From these accounts and from other sources we learn how great the cost of travelling, whether by road or river, used to be. Those who wished to go by road from Calcutta, a distance of 301 miles, had to pay no less than Rs. 406, of which Rs. 301 represented the wages of *palki*-bearers and the balance the hire of the *palki*. Those who went by river had, according to a return for 1781, a journey of 45 days from Calcutta, and its cost depended upon how many *dandis* or boatmen were engaged. For a budgerow the rate varied from Rs. 2 to Rs. 8 per diem, while larger boats called "woollocks", i.e., *utanks*, cost from Rs. 22 to Rs. 29 a day. Steam vessels do not appear to have gone so far up the river until 1828, when a vessel called the *Hooghly* came up from Calcutta; the passage, up and down, was performed at an average speed of 4½ miles an hour.*

The East Indian Railway was extended to Monghyr in February 1862, and since that date several other lines have been constructed. First a line, now known as the Loop Line, was constructed with a great bend to the north, in order to follow the Ganges; and then, when traffic increased, a chord line was made from Lakhisarai to Khana. Twenty-five years ago the construction of the South Bihar Railway continued the line to Gaya on the west, where it meets the Grand

* The Good Old Days of Honourable John Company (reprinted, Calcutta, 1906), Vol. I, p. 488, and Vol. II, pp. 15 and 22.

Chord Line; and more recently the Bengal and North Western Railway extended its system through the north of the district, the line having been opened to traffic in 1900. Since then branch lines have been constructed from Mansi northwards through the Gogri thana into Bhagalpur district, and from Khagaria to Samastipur.

RAILWAYS.

East Indian Railway.

The district is, on the whole, well served by railways, those to the south of the Ganges being broad gauge and those to the north metre gauge. South of the Ganges the Loop Line of the East Indian Railway passes through the district from east to west and the Chord Line from north-west to south-east; while the South Bihar Railway runs through the Sheikhpura thana westwards to Gaya, and there is also a branch line leading from Jamalpur to Monghyr town.

The Loop Line enters the district five miles east of Bariarpur, and leaves it just beyond Barhiya on the western border of the district. This line presents some interesting engineering features. Shortly before reaching Jamalpur, it passes through the northernmost ridge of the Kharagpur Hills by a tunnel called the Monghyr tunnel. This tunnel which till recently was the only one on the East Indian Railway, is 900 feet in length, 23 feet in height and 26 feet in width. After leaving Jamalpur, the line proceeds for about thirty miles close to the Kharagpur Hills, and between Kiul and Lakhisarai crosses the river Kiul by a fine lattice girder bridge of 9 spans of 150 feet each. It then takes a sharp curve in a northerly direction, and four miles further on crosses the Halahar river by another lattice girder bridge of four spans of 150 feet each. In this latter portion the line is laid along an embankment pierced by a number of culverts in order to prevent damage from the floods of the Ganges.

Main Line (Chord Line).

The main line of the East Indian Railway enters the district eight miles south-east of Simultala, and joins the Loop Line at Kiul. The main line, which was known in the earlier days of the railway as the Chord Line, is in this district marked by some steep inclines and passes through picturesque hill scenery. After leaving Simultala, it runs through a pass between the hills, and then through some deep cuttings till it reaches Jhajha, long known as Nawadih. The latter station, which is situated near the hills, contains quarters for the accommodation of the Company's servants, and is a changing place for engines, a second engine being generally attached to goods trains to enable them to ascend the steep incline between this station and Simultala. The line then runs close to the base of the Kharagpur Hills till it reaches Kiul, a large junction, at which the Chord and Loop Lines meet, and from which the South Bihar Railway runs south-west to Gaya.

South Bihar Railway. The latter railway has a length within this district of about twenty-three miles, and after leaving Kiul and Lakhisarai passes two stations, viz., Serari (ten miles from Lakhisarai) and Sheikhpura, six miles further on.

Monghyr branch line. The only other line in the south of the district is the Monghyr branch line, six miles long, which connects Jamalpur and Monghyr, and has an intermediate station at Purabsarai, one of the *mahalas* of Monghyr on the outskirts of the town.

Bengal and North-Western Railway. North of the Ganges the country is served by the Cawnpore-Katihar line of the Bengal and North-Western Railway. Proceeding from west to east it skirts the Ganges fairly closely for some miles from Bachhwara to Begusarai. Thence it continues due east to Sahebpur-Kamal at a greater distance from the river, which takes a sharp bend south before reverting to its easterly course. From Sahebpur-Kamal to Mansi the railway follows the river closely in a northerly bend, and then again assumes a south-easterly direction, roughly parallel to the river. There are short branch lines from Barauni and Sahebpur-Kamal to the Ganges at points opposite Mokameh and Monghyr, the termini being Simaria Ghat and Monghyr Ghat, respectively. Excluding these two short branch lines, the length of the line within the district is about seventy-five miles.

From Mansi there is a branch line, which runs due north through the centre of Gogri thana to Supaul in North Bhagalpur, and thence to Bhaptiahi on the Khanwa Ghat branch line near the Nepal frontier. Another branch line runs from Khagaria through Narhan to Samastipur. Another line to Samastipur, which takes off from the main line at Bachhwara, runs for five miles of its length in Monghyr district. A list of railway stations is given in Appendix I.

ROADS. The district board maintains the main roads, which have an aggregate length of 1,490 miles, of which 159 miles are metalled, besides a number of village roads, chiefly rough unmetalled tracks, with a total length of 121 miles. South of the Ganges the principal roads radiate from three centres, Monghyr, Lakhisarai and Jamui. The old Ganges road, the historical route from Bengal through Rajmahal to Monghyr, and thence to Patna, Benares, Allahabad, and Delhi, runs for fifty-four miles in Monghyr district, from Ghoreghat on the border of Bhagalpur district to Barhiya on the border of Patna district. The fifteen miles between Monghyr and the Bhagalpur border are metalled; there is an inspection bungalow at Bariarpur, eleven miles from Monghyr. Of the western portion, between Monghyr and Barhiya, only the first three miles are metalled. There are ferries at the crossing of the river Kiul at the 29th mile and the Harohar at the 32nd mile from Monghyr; and inspection bungalows at Baha, Surajgarha, Lakhisarai and Barhiya. This is a fine old road, with an abundance of shady road-side trees, ninety or a hundred feet wide, of which in the portion that has been made suitable for modern traffic nine or twelve feet have been metalled.

A metalled road connects Monghyr with Jamui, taking off from the Ganges road at Bariarpur (eleven miles), and passing through Kharagpur (twenty-three miles), Gangta (thirty miles), Gurdih (thirty-seven miles), and Mallepur (forty-six miles) where the Jamui station of the East Indian Railway is situated. At each of these places there are inspection bungalows.

Another good road in the sadr subdivision south of the Ganges is that from Sultanganj railway station to Sangrampur, twenty-three miles in length, of which the first eight miles are metalled. There are inspection bungalows at Tarapur at the twelfth mile and at Sangrampur.

The important roads from Lakhisarai are all second class roads, running to Sikandra (eleven miles), where there is an inspection bungalow; to Sheikhpura, taking off from the Sikandra road at the sixth mile; and to Jamui (eighteen miles), with an inspection bungalow at Majhwe.

There is a partially metalled road from Jamui to Barbigha ($40\frac{3}{4}$ miles), through Sikandra and Sheikhpura, of which the first five miles and the last twelve are metalled. There are inspection bungalows at Sikandra, Girinda, and Barbigha.

North of the Ganges the principal roads are the Tirhut road, the Rusera-Gogri road, and the Nepal road. The Tirhut road runs for forty-three miles, of which $14\frac{3}{4}$ miles are metalled, from Monghyr Ghat through Begusarai and Teghra to Rasidpur. There are inspection bungalows in the 7th mile at Ballia, in the 17th mile at Begusarai, on the 31st mile at Teghra, and in the 38th mile at Bachhwara. The Rusera-Gogri road runs roughly parallel to the Tirhut road from Rusera, through Deraria, Manjhaul, Parihara and Khagaria, to Gogri. The Nepal road, 24¾ miles long in this district, runs from Mansi to Balhi on the Bhagalpur border, whence it continues north to Supaul, Bhaptiahi and Khandauli. There are inspection bungalows at Dhamara in the ninth mile, and at Bakhtiyarpur in the seventeenth mile.

In thana Teghra, and in the western and eastern portions of thana Begusarai, there are numerous branch roads providing fairly easy communication from one part to another and to the railway line; but with the exception of the Nepal road, Gogri thana is practically without means of communication other than surface and fair-weather roads. The nature of the country, the numerous streams, and the liability of the whole of the central area to inundation make the cost of embanked roads prohibitive. The difficulties and expense of road-making in this area may be gathered from the fact that in the first eleven miles of the Nepal road from Mansi even in the driest weather four ferries, and for a great part of the year, six ferries have to be maintained.

WATER COMMUNICATIONS.

The Ganges, which intersects the district from west to east for over seventy miles, is navigable at all seasons of the year for river steamers and the largest country boats. A considerable river-borne trade is carried on, the steamers of the India General and Rivers Steam Navigation Company conveying goods and passengers to a number of places between Calcutta and Patna. The East Indian Railway also maintains a ferry steamer service from Monghyr to the opposite bank of the Ganges. Another steamer service connects Mokameh Ghat on the East Indian Railway with Simaria Ghat on the Bengal and North-Western Railway.

The Little Gandak is navigable for large country boats all the year round, but is only used by river steamers as far as Khagaria, a few miles from its junction with the Ganges. During the dry season the channels at the entrance become unnavigable for river steamers, and dredging is required to allow for their passage throughout the year. The Tiljuga is also navigable for country-boats all the year round, but only small craft of ten tons burden can ply on it during the hot weather. It is not used by river steamers, the number of snags in the river bed for the first few miles being an obstruction to navigation. In 1902-3, the Indian General Steam Navigation Company tried to run steamers up to Dhamahra Ghat from the Ganges at the Kosi Bridge, in order to tap the large grain export traffic of the adjoining country; but the experiment proved a failure. Boats are also largely used as a means of communication in the northern portion of the district, where a large area remains under water during the rains.

CONVEYANCES. The conveyances generally in use, whether boats or carts, are the same as in other parts of Bihar and call for no special description. In the south, however, the people use a small cart of primitive make, called *sagar*, which is characteristic of the uplands of Chota Nagpur. The wheels are solid circles made of *mahua* or *simal* wood joined by a *sal* wood axle, on which are bound a couple of bamboos uniting the wheels to the yoke. This cart is mainly used for the conveyance of timber from the jungle. Where the roads are too stony or steep for the easy passage of carts of this prehistoric type, grain, *mahua* and other produce are mainly carried by pack-bullocks.

POSTAL COMMUNICATIONS. It is of some interest to compare the present state of postal communications with what it was a little over a century ago. From a table of rates of postage issued in 1795 we find that the postage from Calcutta of a letter weighing 2½ tolas was four annas, and heavier letters were charged for at an increasing rate, one rupee being charged for letters weighing 4½ to 5½ tolas. An accident which happened to the *dak* boat in that year shows us how scanty was the correspondence under this system of rates. A *dak* boat containing the Calcutta letters despatched to Bhagalpur and Monghyr having been upset and all the letters lost, a list of both mails was published. The list was not a long one, for there were only four private and four service letters for Bhagalpur, besides a copy of the *Morning Post* and twelve magazines, while for Monghyr there were three private and two service letters and eight magazines only. In striking contrast to this is the fact that in 1924-5 no less than 4,309,724 postal articles were delivered in the district, including 1,198,608 letters, 2,752,299 postcards, 187,806 newspapers, 153,107 packets, and 17,904 parcels.

There are 62 post offices in the district and 171 miles of postal communications. There are 16 combined post and telegraph offices, at Monghyr, Barauni, Barbigha, Barhiya, Begusarai, Gidhaur, Jamalpur, Jamui, Jhajha, Khagaria, Lakhminia, Lakhisarai, Basdeopur, Sheikhpura, Simultala, and Teghra. The value of the money orders issued in 1924-5 was Rs. 30,72,696 and of those paid Rs. 48,87,536. The total amount deposited in the post office savings bank was Rs. 7,28,720 and the number of depositors 9,434.

CHAPTER X.

LAND REVENUE ADMINISTRATION.

REVENUE HISTORY.

UNDER the rule of the Mughal emperors the district appears to have been included in *Sarkars* Hajipur, Tirhut, Bihar and Monghyr. The greater portion was apparently comprised within *Sarkar* Monghyr, which was assessed to Rs. 7,41,000 by Todar Mal in 1582. According to Mr. Grant's account (1787), this *Sarkar* was "altogether or for the most part unsubdued, and probably unexplored, as held by independent or refractory zamindars"; and we may perhaps accept his view that it was only included in the assessment owing to "the ambitious conquering policy of the Moguls, having always in prospect the entire subversion of the lesser as well as the greater states of Hindostan".* However this may be, *Sarkar* Monghyr must have embraced areas not included in the present district, for when the *Diwani* was taken over by the British in 1765, it extended over 8,270 square miles, assessed to a net revenue of Rs. 8,08,000.

The district was constituted in 1832 by the transfer of several *parganas* from the districts of Bhagalpur, Bihar and Tirhut, the land revenue being, it is reported, Rs. 3,82,330 paid by 1,049 estates with 5,583 registered proprietors. Two years later *pargana* Chakai was transferred from the district of Ramgarh, and other changes were made in 1839, 1845 and 1846. At that time land revenue, excise and other revenue were, for the most part, paid into the treasury at Bhagalpur, and the accounts were not kept separately. This continued to be the practice till 1850, when the land revenue of Monghyr was Rs. 7,49,230, the number of estates being 3,581, and of proprietors or co-parceners 26,933. In 1874-5 the number of estates on the revenue roll had increased to 4,053 and the land revenue to Rs. 9,40,340; and it is now nearly the same, the collections in 1923-4 being Rs. 9,96,545.

Owing to the land revenue accounts of Monghyr not having been kept separately before 1850, it is not easy to institute any comparison between the present land revenue

* Fifth Report of the Select Committee (reprinted, Madras, 1883), Vol. I, pp. 507-8.

and the figures for earlier years. It is, however, known that the demand increased largely during the first half of the 19th century as the result of resumption proceedings. At the Permanent Settlement a large proportion of the area was claimed as revenue-free or *jagir* and escaped assessment. In fact, it is estimated that in eight *parganas* of North Monghyr, which were transferred to this district from Tirhut, one-eighth of the area was not assessed. By 1831, however, the resumption proceedings had raised their revenue from one to two lakhs. In *pargana* Pharkiya also the resumption proceedings instituted after the survey of 1835—8 raised the demand from Rs. 46,226 in 1795-6 to Rs. 88,039 in 1846. The demand, which was progressive, has since increased in that *pargana* to Rs. 1,26,238.

SURVEYS AND SETTLEMENTS.

The first professional survey was carried out between 1835 and 1838 in *pargana* Pharkiya by Lieutenant Egerton. This survey was determined upon in order to demarcate certain *wairana* lands, i.e., tracts of waste land outside the ambit of the settled and cultivated villages, to which it was held that the Permanent Settlement did not extend. The survey was confined to boundaries and had neither the accuracy nor the completeness of the subsequent revenue survey. The latter was carried out in the rest of the district by Captain Sherwill in 1845-7, the survey of *pargana* Pharkiya being formally given the dignity of a revenue survey. A survey of *diara* lands subsequently took place in 1865-6; the Srinagar-Banaili estate in North Monghyr, with an area of 174 square miles, was surveyed and settled between 1887 and 1894; and 47 square miles in thanas Teghra and Begusarai were surveyed in 1895-6 in connection with the settlement of the Narhan estate. A record-of-rights has now been prepared for the whole district, for North Monghyr and the Government estates south of the Ganges between 1899 and 1904, and for the rest of South Monghyr between 1905 and 1912.

ESTATES.

According to the Collectorate returns, the number of estates on the revenue roll in 1923-4 was 9,397, including 9,300 permanently-settled estates, 68 temporarily-settled estates and 29 estates held direct by Government; and the current demand of land revenue was Rs. 9,74,079. Owing to the backward condition of the country at the time of the Permanent Settlement, its incidence is low, amounting only to a little over one-tenth of the gross rental of the district.

It is particularly low in North Monghyr, where only a small portion of the area was assessable even as late as 1850; and though a large increase in the demand was obtained in *pargana* Pharkiya in comparatively recent times, that assessment was necessarily low, because even then a large percentage of the area resumed was not under cultivation.

Subdivision of property is known to have gone on rapidly, the number of estates on the revenue roll rising from 4,053 in 1874 to 8,119 in 1907, and to 9,397 in 1924. Apart from these complete partitions, subdivisions into *pattis*, by the Civil Courts or by private arrangement, whereby each group of proprietors hold in severalty with a joint liability for payment of land revenue, has gone to extreme lengths. Thus in North Monghyr, though the total number of estates on the revenue roll was 4,367, the Settlement Officers framed 9,730 separate records of proprietary interests. The number of proprietors was 83,410, and was greatest (21 on the average) in privately-partitioned revenue-paying estate, and least (three on the average) in jointly-held revenue-free properties, many of which are of a petty size.

The area belonging to each proprietor is extremely small, enquiry showing that an average village of 599 acres is ordinarily divided among six different *pattis* with no less than 51 proprietors, and that each proprietor's share is only about 12 acres. In Gogri thana an estate averages 285 acres and each proprietor's interest 69 acres; but in thanas Teghra and Begusarai the estates are exceptionally small, averaging only 40 and 70 acres respectively, while each proprietor's share is 4 and 5 acres respectively. In these two latter thanas alone *khewats*, or records of proprietary interest, had to be prepared for no less than 26,011 estates, 9,831 *pattis* and 68,237 landlords, the smallest recorded subdivision of proprietary rights being $\frac{1}{2,480,000,000}$ of an anna. In a single plot of land, the area of which was just over half an acre, there were 1,582 co-sharers, each of whose shares represented only 7¼ square feet.

Zirat and *Bakasht* land.

Thirteen per cent of the occupied area in North Monghyr, and 11¾ per cent. in South Monghyr, is in cultivating possession of proprietors and tenure-holders. The proportion of land in cultivating possession of properties is highest in Lakhisarai thana (22½ per cent.), where petty landlords keep a considerable part of the best land in their own cultivation.

Proprietors' private land, in the strict sense of the term, is locally known as *kamat;* this privileged land is strictly limited by the provisions of the Bengal Tenancy Act, and only 619 acres in North Monghyr and 311 acres in South Monghyr were so entered in the record-of-rights in the survey and settlement proceedings.

Government estates. There are 97 Government estates in Monghyr, of which 68 are temporarily-settled, while 29 are held under direct management. Most of these estates, with an area of 95 square miles north of the Ganges, and 57 square miles south of the river, were surveyed and settled in the North Monghyr settlement operations. The remainder consist of the Bhaisunda mahals, the Monghyr fort estate, and some petty *mufassal* estates.

The largest class of estates, consisting mainly of *diara* estates, includes those resumed at different dates, but mostly between 1825 and 1840, under Regulation II of 1819. Most of these are *izad mahals*, viz., lands not included in the original settlement through mistake, or subsequent accretions or formations by alluvion. Besides these, there are estates in Gogri, Surajgarha and Sheikhpura which are known as *wairana mahals*. Those in the north of the district were portions of *tappa* Saraunja, which was entirely waste at the time of the Permanent Settlement, and was therefore excluded from settlement. Those in the south were unsettled waste lands lying within the ambit of *parganas* settled with the Raja of Kharagpur, and of which he refused settlement, when it was proposed to resume them. Another group includes estates which were at one time permanently-settled estates in the *diaras*, for which the proprietors took remission of revenue, when it was discovered during the revenue survey that they had diluviated. They were taken possession of by Government, on their subsequent reformation, and in some cases managed direct and in others leased to farmers or to the original proprietors. Ten estates came into the hands of Government at different times by purchase at revenue sales; five others were formed from excess or *amanat* lands out of the area purchased by Government from Buniad Singh, the principal zamindar of *pargana* Pharkiya in the beginning of the 19th century, and from other proprietors, for distribution in the shape of revenue-free grants among the East India Company's pensioned or invalided sepoys. One estate represents land acquired for the East Indian Railway

Company, but nc longer required for railway purposes. The following is an account of the distribution of the estates in the different thanas of the district.

Begusarai thana contains *diara* estates covering an area of over 28 square miles, of which the majority lie within fairly easy reach of Begusarai town, and the remainder are just opposite the town. There are also inland estates grouped round Sisauni, about six miles north of the railway near the Burh Gandak river, which cover, all told, an area of less than 2 square miles. The largest estates are Arazi Bhawanandpur with an area of over 17 square miles, Jafarnagar nearly 5 square miles in extent, and Mahazi Bhawanandpur and Akbarpur Barari covering about 3½ and 2 square miles, respectively.

Gogri thana contains inland estates covering an area of over 20 square miles, and seven *diara* estates, viz., five estates which, as mentioned below, have been recently traced, Tetrabad, which is only 8 acres in extent, and Binda *diara*, which has an area of 43 square miles. The estate last named is now bisected by the Ganges, and for police purposes is divided into two estates, Shumali and Janubi, of which the former is in the jurisdiction of Gogri thana and the latter of Monghyr thana. Of the inland estates, all, except Parbata, which is not far from Binda *diara*, are of considerable size; Cherakhera, Agar and Dhanupra lie some 16 miles north of Khagaria railway station, not far from where the Tiljuga enters the district; Goas and Morasi are even more inaccessible, lying off any good road some 10 miles north of Maheshkund station; Arazi Jalkar Mohani is close to Jamalpur Gogri. These six large estates contain mostly low lands suitable for paddy cultivation, while in the small one, Parbata, high lands predominate and the *bhadoi* and *rabi* harvests are the most important.

In 1908, after the conclusion of the settlement, four petty estates, which were purchased by Government at revenue sales and were long treated as diluviated, were traced and brought under direct management. These estates are called Jagir Raushan Khan Naik Thana Jafra, Jagir Basti Singh Sipahi Thana Jafra, Jagir Manik Singh Sipahi Thana Jafra, and Jagir Gurdayal Naik Thana Jafra. The fifth estate, entitled Dand Sukul Nail Thana Jafra, is still under water.

In thana Monghyr there are Government and temporarily-settled estates with an area of 48 square miles. There are two inland estates, Amanat Sarkar Itahri and Bargoria, which cover barely 120 acres between them, though the lands of the former are scattered over no less than seven villages, some of which are near Bariarpur railway station and others near Jamalpur. Of the *diara* estates, Kutlupur, Tarapur and Zamin Digri (or Decree) are the largest, Kutlupur covering 20 square miles, Tarapur about 13, and Zamin Digri nearly 9 square miles, or 42 square miles in all. Tarapur adjoins Binda *diara*, Zamin Digri is close to Monghyr fort and railway station, while Kutlupur is some 14 miles west, on the border of thanas Monghyr and Surajgarha. A number of petty estates are grouped just opposite or alongside Monghyr town, and the remainder are midway between Monghyr and Kutlupur.

In Surajgarha thana, there are estates covering a little more than 8 square miles; but Rahatpur with an area of 3 square miles, and Kherho Paranpur covering a little over 1 square mile, are the only two of importance. Rahatpur and the petty *diara* estates are all fairly close to Surajgarha, while the inland estates are grouped round Lakhisarai and Kiul.

Of the total number of holdings in Government estates, 68 per cent. have been recorded with rights of occupancy, three-fourths of these being *diara* holdings, occupied for more than 12 years continuously; 31 per cent. have been recorded as non-occupancy, the majority of which are also *diara*. Some difficulty about status arose from the practice, once common in Monghyr, of granting large speculative leases of unculturable *diara* to residents of Monghyr, who had no intention of cultivating for themselves, but took leases at low rates in the hope of being able to make a profit by subleasing later at high rates. The terms of the original lease made it necessary in some cases for the Settlement Officers to record speculative tenure-holders as raiyats, and this accounts for the number of under-raiyati tenancies recorded.

Bhaisunda *Mahals.*

The estates known as the Bhaisunda *mahals* are a curious survival of the early settlement of *pargana* Pharkiya. At the time of the decennial settlement, revenue was specially and separately assessed on the assets derivable from grazing fees. This assessment was called *bhaisunda* and continued to be made even after the Permanent Settlement. When resumption proceedings were started in the *pargana,* and the land settlement was made permanent, the *bhaisunda* settlement still continued to be temporary. The result of this arrangement, of the numerous changes in ownership which have taken place during the period that has since elapsed, of the enormous increase in the cultivated area and the corresponding decrease in the area suitable for grazing, is that in recent years there have been constant defaults of the settlement holders of the Bhaisunda *mahals*, who are in most cases not the proprietors of the villages in which those *mahals* lie. Of late years it has been difficult to get any one to take up their settlement, and it has consequently been proposed by the Settlement Officer that these *mahals* should be struck off the revenue roll as the current terms of settlement expire.

TENURE-HOLDERS. *Ghatwali* tenures.

In South Monghyr a number of estates were formerly held on the *ghatwali* tenure, especially in the Kharagpur estate. The origin and nature of these tenures have been described as follows in a judgment delivered by the Privy Council in 1855.* "The mountain or hill districts in India were inhabited by lawless tribes, asserting a wild independence, often of a different race and different religion from the inhabitants of the plains, who were frequently subjected to marauding expeditions by their more warlike neighbours. To prevent these incursions it was necessary to guard and watch the *ghats*, or mountain passes, through which these hostile descents were made; and the Muhammadan rulers established a tenure, called *ghatwali* tenure, by which lands were granted to individuals, often of high rank, at a low rent, or without rent, on condition of their performing these duties and protecting and preserving order in the neighbouring districts. Nothing could be more deplorable than the state of the provinces under this system. Murder and rapine were common throughout the country; more than half the lands were waste and uncultivated; and neither the raiyats nor the zamindars had any inducement to improve them, as any increase in their value had only the effect of increasing the Government assessment.

"It was considered by the East India Company that the first step towards a better system of government and the amelioration of the condition of their subjects would be to

* *Lilanand Singh* v. *Government of Bengal,* Moore's Indian Appeals, Vol. VI, p. 101, *et seq.*

convert the zamindars into landowners, and to fix a permanent annual *jama*, or assessment to the Government, according to the existing value, so as to leave to the land proprietors the benefit of all subsequent improvements. Accordingly, they determined to make the assessment in the first instance for a period of ten years, with a view to its being ultimately made permanent." After describing the decennial settlement, the judgment goes on to say that at this time Raja Kadir Ali was the zamindar of Kharagpur, " a considerable principality including many *parganas*," and that a very large quantity of land had been granted by his ancestors on the *ghatwali* tenure.

" The extent and particulars of these vast estates, and the nature of the *ghatwali* tenures, were well known to the Government of Bengal at the time when the settlement was made. Some years before, in consequence of disturbances which had taken place in the country during the time of Kadir Ali's father, the Government had found it necessary to interfere with a military force, and having displaced the then Raja and restored tranquillity, had placed the zamindari under the charge of one of their own officers, Mr. Augustus Cleveland, who had the management of it up to the year 1781, about which time Kadir Ali (his father having died) was put into possession of the Raj. It appears that Mr. Cleveland, during the time that he was in charge of these estates, had granted no less than 87,084 *bighas* of land upon *ghatwali* tenure, in conformity with the orders of Government. It appears from other evidence that the grants before Mr. Cleveland's time to the *ghatwals* reserved a payment of two annas per *bigha* as a fee or perquisite to the zamindar; that some *sanads* were granted unadvisedly by Mr. Cleveland without such reservation, but that he afterwards insisted on such payment being made to the Government while he was in charge on behalf of the Government, and that all grants subsequently made by the Raja of Kharagpur contained the same reservation.

" In 1813 a report was made by the Collector of Bhagalpur to the Magistrate of Birbhum, in answer to certain enquiries with respect to *ghatwali* lands in his district. The Collector states that the *ghatwali* lands in his district are of four kinds. First, the lands already referred to as granted to Mr. Cleveland. These he states to have been allotted in the environs of the forests, at the foot of certain mountains ' to certain *ghatwals* and watchmen, in lieu of salaries, to attend to and guard the watch stations at the passes, and to patrol the precincts of the villages, that no mountaineers might be able to descend from those passes of the mountains to commit night attacks, to invade or assault, or to plunder money or cattle, or to create disturbance'. The second class the report describes as ' the *ghatwals* attached to the Kharagpur estates, who pay a stipulated rate of rent of their lands and villages being bound to protect and guard the highways, to watch the stations at the passes, to prevent disturbances being created by the mountaineers, thieves and highwaymen. They hold their lands in virtue of *sanads* granted by the zamindar of Kharagpur except some who have received theirs from the former authorities '.

" The report then proceeds to state that when the zamindar, or Government authority, wishes to appoint a *ghatwal* to guard the frontiers of the villages, it is his duty to ascertain the produce of the villages, the quantity of *ghatwali* lands therein, and, after deducting a certain rate in the ratio of the guards with the *ghatwals*, in lieu of wages, to fix a certain rent to be paid by the *ghatwals*. After mentioning other descriptions of *ghatwali* lands, he states his opinion that the *ghatwals* have no right of inheritance or proprietary interest in their lands, but hold right of possession as long as they perform the terms and conditions of their *sanads*. The report then states that at the time of the decennial settlement the *ghatwals* were not treated as independent *talukadars;* that no settlement was made with them, but that they were included in the settlement of the zamindar of whom their lands were held. In 1816, another report was made by the Collector of Bhagalpur, in which it is stated that the *ghatwals* pay a fixed rent to the zamindar of Kharagpur, and continue under his control, direction and subjection, while the Raja is answerable to the Collector for the rents of the entire district of Kharagpur."

Under the provisions of the decennial settlement, the Bengal Government, in 1790, assessed the whole of the zamindari of Kharagpur, including *ghatwali* lands, at a fixed *jama*. This settlement was made perpetual in 1796, under Bengal Regulation I of 1793, at the same fixed *jama*. In 1838, the Government set up a claim to resume the *ghatwali* lands for the purpose of revenue assessment ; but the claim was dismissed, on the grounds that (1) the *ghatwali* lands were part of the zamindari of Kharagpur, were included in the permanent settlement of the zamindari, and were covered by the *jama* assessed on the zamindari ; and (2) lands held under *ghatwali* tenure were not liable to resumption under Regulation I of 1793.

At the time this suit was instituted, the Kharagpur estate was still in possession of the Raja of Kharagpur, but it was sold up for arrears of land revenue in 1840 and purchased by Raja Bidyanand Singh, the grandfather of the present proprietors of the Banaili estate, and by Balnath Sahu, who next year transferred his interest to Bidyanand Singh. On his death, the suit was carried up to the Privy Council by his son and heir Raja Lilanand Singh. It was then decided that the Kharagpur *ghatwali* tenures are perpetual and hereditary grants of land, which cannot be resumed by Government. After this decision, the *ghatwali* tenures were restored and the Raja instituted suits to resume. Government, however, ruled in 1863 that the *ghatwali* services were still demanded from him, and that, so long as Government demanded them, he could not resume the tenures. The Raja thereupon agreed to pay Rs. 10,000 a year in lieu of the services for which he was responsible, he being left to make what arrangements the Courts would allow his *ghatwals*. When, however, he instituted suits in 1864, the Courts decided that, although Government had dispensed with the *ghatwals'* services, the lands could not be resumed, and that the *ghatwals* had permanent hereditary tenures at a fixed *jama* and could not be evicted except for misconduct.* On the other hand, it was held in 1866 in another case, on a consideration of the terms of a lease under which a *ghatwal* held his lands, that the zamindar could resume them when the *ghatwali* services were no longer required.†

There have been numerous other cases regarding the *ghatwali* tenures of Kharagpur, in which their legal incidents have been laid down. In one case it was decided that the lands of the *ghatwals* are not capable of alienation by private sale or otherwise, are not liable to sale in execution of decrees, except with consent of the zamindar and his approval of the purchaser as a substitute for the out-going *ghatwal*.‡ In another case it was held (1) that a *ghatwali* tenure in Kharagpur is transferable, if the zamindar assents

* *Manranjan Singh* v. *Lilanand Singh* (3 W. R. 64).
† *Lilanand Singh* v. *Sarwan Singh* (5 W. R. 292).
‡ *Lilanand Singh* v. *Durgabati*, W. R. Sp. No. 1864, 249; *Guman Singh* v. *Grant*, 11 W. R., 292.

and accepts the transfer, which assent and acceptance may be presumed from the fact of the zamindar having made no objections to a transfer for a period of over 12 years, and (2) that in dealing with a *ghatwali* tenure the Court must have regard to the nature of the tenure itself, and to the rules of law laid down in regard to such tenures, and not to any particular school of law or the customs of any particular family, inasmuch as a *ghatwali*, being created for a specific purpose, has its own particular incidents and cannot be subject to any system of law affecting only a particular class or family. In the latter case it was pointed out that there is this difference between the *ghatwals* of Birbhum and those of Kharagpur that the former are appointed by Government, and the latter by the zamindar.*

Most of the *ghatwali* tenures of Kharagpur have now been sold up by the proprietors of the Banaili Raj and let out on *mukarari* leases. A portion of village Mangrar in *pargana* Parbatpara is still held in *mukarari* by the descendants of the former *ghatwals*, who retain the title of Thakur, but the other *mukararidars* are outsiders, lawyers of Bhagalpur, Baniyas of Jamui, etc.

Chakai *ghatwali* tenures.

Originally the estates in *pargana* Chakai to the south were similarly held on *ghatwali* tenure. About 1774 the lawless state of this tract led the British to place it in charge of Captain James Browne, who settled the estates with the *ghatwals* with two exceptions. These two exceptions were Dumri and Mahesri, which were settled directly with the proprietors, the story being that the *ghatwal* tenure-holders fled at the approach of Captain Browne, their reputation as dacoits and brigands being too strong for them to face a Government officer without fear of the consequences. In the case of Dumri, however, the *ghatwals*, finding that in their absence a settlement had been made of their tenure, returned and obtained a *sanad* settling it with them under the Raja of Gidhaur. Of the estates settled with *ghatwals*, only two are now held by their descendants, viz., Tilwa and Kewal. The others have passed into the hands of the Maharaja of Gidhaur, Chetru Rai, Akleswar Prasad and others of Rohini.

Ghatwali mukararis.

The ruin of the *ghatwals* of Chakai is attributed to their improvident system of management and to their family customs. Generally, the eldest male member in the nearest

---

* *Ananda Rai* v. *Kali Prasad Singh* (I. L. R. 10 Calc. 677; S. C. 15 Calc. 471).

line of descent succeeds to the *gadi*, as it is called, though occasionally the widow of a proprietor has been allowed to hold as Thakurain. Other members of the family are provided for by *mukarari* grants called *babuana*, which are generally inherited by the descendants of the original grantee, though they originally were meant for the support of the grantee during his life-time only: a *mukarari* which terminates with the life of the grantee is known as *hinhayati*. In addition to *mukararis* granted to members of the family, the *ghatwals* used to lease out large areas of jungle land in perpetual *mukarari*, at a nominal rental, for the extension of cultivation. Numerous rent-free grants were also made as service and religious tenures on the slightest grounds. The proprietors' rent-roll thus stood little chance of increasing, while their expenditure was in excess of their income. They consequently got deeply into debt, and had to mortgage their estates, which were subsequently sold up in liquidation of the debts and purchased by the mortgagees. The *mukarari* tenures granted by them have in most cases passed into the hands of the new landlords, either by purchase or by forcible resumption, and in Chakai not many have survived.

Temporary tenures.

The *zarpeshgi thikadar*, an usufructuary mortgagee, who holds his tenure as security for the mortgage loan, collecting rent in lieu of interest, is common throughout the district.

A variety of this tenure is the *satua patua* lease, under which both principal and interest are liquidated by the realisations of the mortgagees. The *mustajiri* tenure, which is common in the Jamui subdivision, calls for more detailed description. In this part of the district few villages are held directly under the zamindar. The majority are held by *thikadars* or *mustajirs*, who, in the more highly cultivated villages of Jamui thana, are mere farmers of rent. They make their profit, for the most part, out of the cultivation of lands which they hold by virtue of their position as *thikadars;* but in the more or less jungly villages, which make up the greater part of the subdivision, they have a status which it is often difficult to define. In the course of the settlement proceedings, these *thikadars* have been divided into three classes. (1) First, there is the ordinary farmer, an outsider pure and simple, who takes on *thika* a village with which he has no previous connection, with the object of collecting rents and making a profit out of the margin between his collections and the amount payable under the lease. Persons of this class are recorded as *ijaradars*. (2) There is next the case where the proprietor has given a *thika* lease to a person who, prior to the execution of the lease, was a raiyat of the village and had occupancy rights in his lands. He is also recorded as an *ijaradar*, but the lands in which he had previously acquired occupancy rights are recorded as his occupancy lands. (3) The third kind of *thikadar* is the person who originally received from the proprietor a reclamation lease for the village, *tola* or *chak* in question. In some instances the original lease is still extant, and is called a *chakband* lease. It defines by boundaries the area within which the lessee has the right to reclaim and the original rent fixed. In a few cases that rent has not been changed to the present day, but in the majority of cases there have been frequent enhancements of rent. Such enhancements have often been accompanied by the execution of new *thika* leases or *miadi kabuliyats* for periods usually of seven years. The original lease is not producible in many cases, but from the history and circumstances of the village or *chak* it may generally be inferred without any doubt that such a lease did exist, or that the predecessor in interest of the present *thikadar* began his connection with the tenancy on a verbal agreement which had the same effect as a *chakband* lease. These cases were dealt with under the relevant provisions of the Tenancy Act. For instance, if the present holder could show that the lands within the tenancy now in his own occupation were reclaimed by himself or by his predecessors in interest, the presumption would be that he was a raiyat for the whole area. If the original lessee at once sublet the whole or most of it to other persons introduced by him for purposes of reclamation, and the lands, if any, held by him were originally reclaimed by others, it may be presumed that the present lessee is a tenure-holder. In the great majority of cases, however, the terms of the Bengal Tenancy Act, read in the light of the facts, necessitate the conclusion that the tenancy is raiyati.

The following account of the *thika* system is extracted from a note contributed by the Settlement Officer. "The system originated in the grant of a tract of jungle land for the purpose of reclamation to some raiyat, who generally belonged to one of the aboriginal tribes, such as the Santals, Naiyas or Bhuiyas. The first lease was granted on a very low rent to be paid for a term of years, generally seven years.

The original lessee, with the members of his family and some fellow-castemen as partners, built their huts on the land, and subsisting at first mainly on *mahua* and other produce of the jungle, commenced to clear the forest, dam up streams, and carve their beds and banks into fields. With the gradual increase of cultivation the rent was enhanced at the end of every septennial period, not on any accurate computation of the amount of cultivation, but probably after a certain amount of haggling between landlord and tenant as to what the latter could now afford to pay.

"The original reclaimer and his descendants have in many cases disappeared from the village, and a new *thikadar*, who was brought in as a cultivator by the first, reigns in his place. He still lets out for cultivation such lands as remain fit for reclamation, and generally conducts the agricultural administration of the village; but the principal object of his existence is no longer to turn the jungle into fields for his own profit and that of the zamindar, but to act as a medium between the zamindar and the raiyats of the *tola*, whereby the former may be able to realize his rents with the least possible difficulty. The zamindar deals only with the *thikadar*, on whose shoulders falls all the burden, if he cannot realize from the raiyats the full amount of their rent. Where the *thikadar* is still merely an ignorant raiyat of the village promoted to be rent collector, he frequently has no voice in the assessment of rents on the various holdings in the village. His own rent is increased every five or seven years according to his agreement, generally verbal, with the proprietor, and he gets a *hukumnama* from the proprietor to increase the rents of the raiyats by a certain proportion so as to make up the increase in his *thika* rent. In many *tolas*, where there is no longer a scope for reclamation, the *thikadar* is an outsider, frequently a servant of the zamindar, who is allowed a small profit on the amount which he realizes from the raiyats. The same septennial increase is taken in these cases also; but sometimes a *tola* is met with where the limit of increase has long ago been reached.

"Where the *thika* system exists for actual reclamation purposes, it is, or would be, if properly managed, the best possible. The *thikadar* is directly interested in developing the agricultural resources of his village as fast and as far as possible, because he gets for his own profit the rent that he can realize from the lands cultivated within each period of settlement.

Even in this case, however, the system is liable to abuse through careless management, for the landlord takes no accurate account, at the end of each period of lease, of the amount of land actually brought under cultivation during the period, but increases the rent by mere guess-work. The result is that the rent goes on increasing arbitrarily even long after the maximum of cultivation is reached. If cultivators are Santals, they stand this up to a certain point, till they think that their profits are less than they can make in some new area, when they depart, leaving the results of their labours to Goalas, Babhans, Modis and others. Where the limit of reclamation has been reached, as far as is possible with the means and capital at the disposal of the *thikadar* and raiyat, or where the settlement is one for the farming of the rents merely, the *thika* system is bad in every way. The periodical demands for increased rent lead to bad feeling between the proprietor and tenants. The burden of the increase tends to fall more and more on the low caste raiyats, who cannot resist the *thikadar's* demands, as the higher caste Babhans and Rajputs do. In consequence, the incidence of the rents on these tenants becomes so great, that the failure of the crops in any year drives them to borrowing first on the security of their cattle and household goods, then on that of their holdings, which pass in a few years into the hands of the Modis and Babhans, who have superior resisting power and outside sources of income.

" Further, in its extremest form, the *thika* system is a system of bleeding. The temporary farmer cares for nothing but to get as much profit as he can out of the village during the term of his lease. He will not make any outlay on improvements because his period is too short for him to expect any adequate return; and he cannot be at all certain that he will be able to get the lease for a succeeding term. The raiyats themselves have to pay rents too high to allow them to expend any money even on the upkeep of such irrigation works as exist, and the *thikadar* will not assist them. Consequently, not only is there no development of the resources of the village by formation of irrigation works, which are absolutely necessary for the security of crops in the area; but such tanks and *ahars* as do exist are allowed to fall into disrepair and to silt up. In fact, the *thikadar* prefers that they should be silt up, because he can then annex their beds to his own *bakasht* land and grow excellent *rabi* crops of wheat and barley on them. The practical result of the purely farming system is thus to decrease the quantity and quality of cultivation in the village rather than to increase it, and to throw a more and more heavy burden on the shoulders of the raiyats, whose rents increase while their crops diminish, and who have no one to whom they can turn for assistance."

Rent-free tenures.

The largest revenue-free tenure in the district is the Abhaipur *pargana* held by the *khankah* or monastery at Maulanagar, the grant of which was confirmed by a *sanad* of Council, dated 9th February 1786. The proceeds of the estate are intended for feeding travellers and beggars, keeping up a school and mosque, and also for the personal expenses of the endowed family. Generally, however, rent-free tenures in Monghyr are of a petty nature and of several kinds, such as *sivottar, brahmottar, devottar, bhatottar, jagir, baksh* or *baksh lakhiraj*. In the Chakai *pargana* a number of these tenures were created by *ghatwals*, often on trifling grounds. For instance, a Brahman recited a chapter of the Ramlila on an auspicious Tuesday and he was given 2 *bighas* of rice land as *brahmottar;* another acted as priest in a *Santnarain katha* and was given 3 *bighas*. Such grants are mostly of recent origin, the *sanads* being from 30 to 60 years old. Some of the latter contain terrible imprecations on any of the successors of the donor who may interfere with the grant. *Bhatottar* tenures were similarly created by the *ghatwal*, each of whom had his *bhat* or *jongleur*. These men, who seem to have a hereditary genius for composing extempore adulatory verses, served as the chroniclers of the *ghatwals*' genealogy. Many *jagirs* or service tenures also owed their origin to the *ghatwal*, who hardly ever paid anybody in cash, but had his drummer, his bugler, his potter, his carpenter, his *paik* and his *barkandaz*, his barber and his *dhobi*, as well as his hereditary priest. All these were paid in *jagir*, and most of the *jagirs* have been left untouched by the present proprietors.

Baksh.

Baksh grants are another curious kind of tenure, which was common in the *ghatwali* estates. It is reported that it was considered necessary for the dignity of a *ghatwal* chief to keep a number of mistresses, and their children were generally maintained by means of rent-free *baksh* grants, *baksh* being the root of the Persian verb *bakshidan*, 'to grant'. Some of these illegitimate children exercised considerable influence, and several *sanads* by which they made *brahmottar* and *sivottar* grants are still in existence. Again, *baksh* grants were made to the *patwaris*, *diwans*, and other Kayasth employés of the *ghatwal*. For instance, a Kayasth of Kiwa was engaged as tutor to a son of a *ghatwal*. After 2½ months it was found out that the boy had become wondrously learned, and a *baksh* of 40 *bighas* of land was immediately given to the tutor.

RAIYATS' HOLDINGS.

The great mass of the tenants are settled and occupancy raiyats. The number of holdings of this class is 550,083; and 1,384,548 acres, over 82 per cent. of the occupied area, is held by tenants with occupancy rights. Raiyats at fixed rates hold less than one per cent. of the occupied area, 5,361 such tenants holding 11,442 acres. 14,370 non-occupancy raiyats hold 56,182 acres; 20,419 rent-free raiyats hold 17,894 acres; and 35,000 under-raiyats are in occupation of 41,049 acres.

Chakband system.

In the Jamui subdivision a number of raiyats hold under the *chakband* system, which is analogous to the system of *thika* leases granted for the purpose of reclamation, a *chak* of land being leased out to whoever wished to reclaim it on a lump rental for a term of years. At the time of the revenue survey of 1847, it may be explained, the jungle was much more continuous and extensive than it is now, and large tracts of land forming one estate, but comprising numerous *bastis* scattered throughout the jungle, were surveyed as one village. These revenue survey villages are now divided up into numerous villages, known locally as *mauzas*, *tolas* and *kitas*. The *mauza* is the traditional site of the original jungly village, and the *tolas* and *kitas* are more recently reclaimed portions of the jungle, which were recognized as appertaining to the old *mauza*. The areas included in these *tolas* and *kitas* were originally granted in *thika* lease for reclamation purposes to various raiyats. It frequently happened, however, that the original reclaiming *thikadar*, finding that he had more land than he could reclaim himself, and not wishing to take the trouble of actively supervising the process of reclamation as carried on by his partners and under-raiyats, would grant to one or more raiyats portions of the village, defined by certain boundaries, as *chakband* holdings. These were generally not larger than could be conveniently reclaimed by one family; and consequently they have preserved their original character intact. Sometimes, however, the *chakbands* were large areas, beyond the power of one family to cultivate, which were, therefore, developed by the help of co-sharers and under-raiyats.

Such comparatively large *chakband* leases are generally known as *katkana* leases, which simply implies that the lessees are subordinate to the *thikadar*. It is reported that in the villages of *taluk* Mallepur, in *pargana* Parbatpara, belonging to the Banaili Raj, these *chakbands* have come to imply a species of *mukarari* interest vested in the original settler and his descendants. This is said to be accidental, being due apparently to the fact that the Raj has not increased the rents of its tenants for several generations; and in no other part of the subdivision do these *chakbands* imply any special rights as regards fixity of rent or security of tenure beyond the ordinary incidents of a raiyati holding. A *chakband* in Mallepur can change hands in its entirety by sale, provided the purchaser obtains the permission of the Raj; but when it comes into the hands of the *malik* through sale for arrears of rent, its *chakband* character is broken, and it is measured and settled anew as *kurokar* land, i.e., measured, from *kuroh* meaning a *bigha*. Elsewhere in the subdivision, in parganas Gidhaur and Chakai, an ordinary raiyati *jot*, as distinguished from a *chakband*, is *phutkar jot*, i.e., broken or composed of scattered plots.

The history of the way in which rents are assessed on such holdings is interesting. Cases have been met with in the *ghatwali* villages of Chakai which show that the rent first fixed for many of the *chakbands* was to be paid in perpetuity, but scarcely any of these *mukarari* leases have survived the efforts of the more modern proprietors to increase their rent-roll, as the cultivation of their estates increases in extent. The great majority were leased out with an implied condition that the rent would be increased at the expiry of the term fixed in the *patta*, if the cultivation within the *chak* had extended to a degree sufficient to bear the increase. In assessing the new rent at the end of the term of lease, however, no proper estimate of the capabilities of the land reclaimed within the period was or is made. If the amount fixed left the lessee what he considered a fair amount to live on, he acquiesced. If he thought it was too high, he protested and generally got the amount reduced. The proprietor could not afford to be too high-handed in his assessment, while there was much jungle land lying ready for reclamation; for the raiyat, who was generally an independent aboriginal, would think it little hardship to leave a zamindari where he was too harshly dealt with, and pitch his hut in the forest territory of some more lenient landlord. This independence of the Santal and other aboriginal tribes is one reason why they give way, after reclamation has reached a certain pitch, to Hindu cultivators, who will cheerfully pay a much higher rent than the Santal and still keep clear of debt. It is probable that at one time the whole area was held on such progressive lump rentals. Even in the highly cultivated villages near Jamui, *chaks* still exist which have come down in the family of one raiyat for generations; but generally they have long ago become *phutkar jots*, and rent is paid, or is claimed, on a classification of the fields.

Alagi jots.

A form of holding arising from the comparative infertility of the *tanr* lands of the southern area is known as *alagi jama*, i.e., a separate *jama*. This consists of land outside the proper holding of the raiyat, which he has a right, obtained from the proprietor, to cultivate on condition of paying a rent at the rate of four to ten annas per *bigha*, only for the year in which he cultivates them. These lands produce a scanty crop of *kodo* or *kulthi* once every three or four years, and are allowed to lie fallow, bearing no rent for two or three years between each crop.

Thika arazi kamat.

Another form of tenure which some proprietors have endeavoured to introduce contrary to the provisions of the Bengal Tenancy Act, is locally known as *thika arazi kamat*. This consists of land formerly held by raiyats which has come into the proprietor's hands either by the departure of the cultivator or by purchase in a sale for arrears of rent. This land is resettled, either with new raiyats or with the *thikadar* of the village, for a term of 7 to 11 years, the lease specifying that it is *kamat* land which the lessee must cultivate himself and never sublet; and that the lessee is to have no rights to the land at the end of the period of lease beyond what may be granted by the proprietor on a renewal of the *patta*. It was generally found during the course of the settlement proceedings that where such lands have been settled with raiyats, they have occupancy rights in them, either having continued to cultivate for over 12 years or being originally settled raiyats of the village; that where such lands have been settled with *thikadars*, they have been unable to cultivate them themselves and have sublet them to raiyats without the knowledge of the proprietor; and that these raiyats have also in general acquired occupancy rights over the land.

CHAPTER XI.

GENERAL ADMINISTRATION.

THE LEGISLATIVE COUNCIL. MONGHYR district returns four members to the Legislative Council, one representing the Muhammadan and three the non-Muhammadan rural electorate, apart from the share which the landholders and the urban population take in returning their members for the Bhagalpur Division. Of the urban electorate of the Division Muhammadans return one member and non-Muhammadans two, while one member represents the landholders of the Division.

ADMINISTRATIVE CHARGES AND STAFF. For administrative purposes the district is divided into three subdivisions, of Monghyr, Jamui and Begusarai, with an area of 1,915, 1,303 and 750 square miles respectively. At Monghyr the Collector is assisted by a staff of seven Deputy Collectors, and by two Sub-Deputy Collectors. The subdivisional officers of Jamui and Begusarai are each assisted by a Sub-Deputy Collector.

REVENUE. The revenue of the district under the main heads rose from Rs. 15,53,500 in 1880-1, to Rs. 17,71,000 in 1890-1, to Rs. 20,70,000 in 1900-1, and to Rs. 23,03,000 in 1907-8, when the first edition of the Gazetteer was in preparation. In 1923-4 it amounted to Rs. 34,71,000, of which Rs. 9,96,545 (including collection of arrears amounting to Rs. 70,512) were derived from land revenue, Rs. 10,52,344 from excise, Rs. 6,86,901 from stamps, Rs. 5,07,981 from cesses and Rs. 2,27,322 from income-tax.

Land revenue. The collections of land revenue aggregated Rs. 8,87,000 in 1880-1, and Rs. 9,32,000 in 1907-8. The current demand in 1923-4 was Rs. 9,74,079 payable by 9,397 estates, Rs. 7,77,837 being due from 9,300 permanently-settled estates, Rs. 54,160 from 68 temporarily-settled estates, and Rs. 1,42,082 from 29 estates held direct by Government.

Excise. The excise revenue increased from Rs. 3,38,988 in 1892-3 to Rs. 5,89,000 in 1907-8, and to Rs. 10,52,344 in 1923-4. The greater portion of the excise revenue is derived from the tax on country spirit prepared by distillation from molasses and the flower of the *mahua* tree (*bassia latifolia*). The receipts from this source in 1923-4 amounted to Rs. 5,37,636, the consumption being three gallons per hundred of the total population. The consumption of the fermented liquor known as *tari* is also considerable, and in 1923-4 a revenue of Rs. 1,14,384 was derived from this source.

The receipts from hemp drugs and opium account for practically all the remainder of the excise revenue. The greater part of the revenue they yield is derived from the duty and license fees levied on *ganja*, the dried flowering tops of the cultivated female hemp plant (*cannabis sativa*). The consumption of hemp drugs is in fact unusually great, the receipts in 1923-4 being Rs. 3,97,215, representing a consumption of 29 seers per 10,000 of the population. On the other hand, opium is not much used; in 1923-4 the duty and license fees on this drug brought in Rs. 44,247, and the incidence of consumption was not more than three seers per 10,000 of the population.

Stamps

The revenue from stamps ranks next in importance as a source of income to that derived from excise. The receipts from this source increased from Rs. 3,21,000 in 1897-8 to Rs. 4,21,000 in 1907-8 and to Rs. 7,35,605 in 1924-5. The increase is mainly due to the growing demand for judicial stamps, which brought in Rs. 5,91,170 as against Rs. 3,41,000 in 1907-8. The revenue from non-judicial stamps, which in 1907-8 was about Rs. 80,000, had risen by 1924-5 to Rs. 1,44,435.

Cesses.

Road and public works cesses are as usual levied at the rate of one anna in the rupee. The current demand in 1923-4 was Rs. 5,35,000, the greater part of which (Rs. 5,17,693) was payable by revenue-paying estates.

Income-tax.

In 1924-5 the net demand on account of income-tax was Rs. 1,21,913 and of super-tax Rs. 11,614. There were 888 persons assessed to the tax, and three persons, whose principal business was money-lending were assessed to super-tax. The Peninsular Tobacco Company was formerly the biggest assessee, but on the reconstitution of the Company its head office was located in Calcutta, and its assessment has consequently been transferred to Bengal. The majority of the persons assessed to income-tax in the district are money-lenders and traders in grain and cloth.

REGISTRATION.

There are eight sub-registry offices, including the one at sadr, under the Indian Registration Act XVI of 1908. At the headquarters station (Monghyr) the District Sub-Registrar deals as usual, with the documents presented there, and assists the District Officer, who is *ex-officio* District Registrar, in supervising the work of Sub-Registrars who are in charge of the other registration offices. The average number of documents registered annually during the quinquennium ending in 1924 was 28,377 as against 25,259 in the preceding five years, there being an increase of about 12 per cent. which is attributed to bad crops. The marginal statement shows the number of documents registered and the receipts at each office in 1924.

—	Documents registered.	Receipts.
		Rs.
Monghyr ...	3,660	24,016
Lakhisarai ...	2,761	10,072
Jamalpur Gogri ...	6,122	12,887
Kharagpur ...	2,580	6,736
Sheikhpura ...	1,776	7,389
Begusarai ...	5,049	14,272
Teghra ...	4,066	10,272
Jamui ...	3,697	8,759
Total ...	29,711	94,403

There are five offices for the voluntary registration of Muhammadan marriages and divorces under Act I (B.C.) of 1876. The Muhammadan Registrars appointed under the Act to register such marriages and divorces also hold the office of Qazi under Act XII of 1880. The average number of registrations annually between 1919 and 1924 was 27, as against 18 in the preceding five years. The headquarters of the Muhammadan Registrars and Qazis are at Monghyr, Begusarai, Sheikhpura, Dahgana (Gogri), and Lakhisarai.

ADMINISTRATION OF JUSTICE.

Civil Justice.

Monghyr was formerly under the jurisdiction of the District and Sessions Judge of Bhagalpur; but since 1914 it has had a District and Sessions Judge of its own. He is ordinarily assisted by two Subordinate Judges and five Munsifs. Two Munsifs are stationed at Monghyr, two at Begusarai, and one at Jamui.

Criminal Justice.

The sanctioned staff of magistrates at Monghyr consists, in addition to the District Magistrate, of seven magistrates of the first class and two of the second or third class. In addition to these officers two sub-deputy magistrates exercising second or third class powers are usually posted there. The Subdivisional Magistrates at Jamui and Begusarai are officers vested with first class powers, each assisted by a sub-deputy magistrate with second class powers. There are Benches of Honorary Magistrates at Monghyr, Jamui, Begusarai, Khagaria, Jamalpur and Lakhisarai.

CRIME.

Regarding the nature of the crime committed in the district the following remarks of a former Commissioner of the Division may be quoted :—" South Monghyr is, I think, the most criminal area which I have seen in India ; and one whole *pargana* is, and long has been, reputed to have a population of thieves. It is in South Monghyr that the old Hindus and Muhammadans and the semi-Hinduized or Islamized people are in strongest contrast, with little inter-dependence, with contempt and repulsion on one side and little reverence or fear on the other; and these social conditions appear to me to account for the comparatively excessive criminality that prevails." These remarks need some qualification; the District Magistrate reports that there has been considerable improvement during recent years in the south Monghyr area, due in part to suppression of criminal gangs, and in part to a movement for reform among the unruly Dosadhs and Dharhis themselves.

CRIMINAL CASTES.

Banpars.

The following is a brief account of the Banpars and Dharhis : an account of the Chakai Dosadhs has already been given in Chapter III. The Banpars, or as they are also called the Banpar Mallahs, are a caste of boatmen and fishermen, who are also skilful sportsmen, catching the *gharial* and crocodile in strong rope nets and eating their flesh. To these innocent occupations they add that of the professional river dacoit. They go far afield in pursuit of the latter exciting profession, and a gang composed of Banpars from Patna and Monghyr has been discovered at Salkia in the Howrah district. In the Monghyr district they number about 1,200 of whom nearly half reside in the Monghyr and Gogri thanas.

Dharhis.

The Dharhis are well-known thieves and dacoits, but, shrewdly enough, they rarely commit crime in the vicinity of their own villages. The offences for which they have been mostly convicted are committed on land, and their connection with the rivers is of a peculiar nature. They do not ordinarily follow any calling on the rivers, but when the country is flooded, they take advantage of it to organize regular raids in little dug-outs, which are kept carefully sunk and hidden away. It is said that when on expeditions outside the district they work in gangs of seven men. They usually live in a separate quarter of the village, and their houses have nearly always a pig-sty attached to them, to which they give the name *benkor*. In the courtyard of their houses there is a small *pindi* dedicated to Ram Thakur, on which they sacrifice the animals they breed, at the same time offering incense and prayer to the god. They have a priest (*bhagat*), who officiates for them at various domestic ceremonies, and also on special occasions, when the Dharhis set out on a thieving expedition.

In appearance the Dharhi is not unlike the Musahar, but he is better developed and usually a more powerful man. Generally speaking, they are a hardy set of men of good physique capable of travelling long distances at a high rate of speed. Out of about 200 prisoners medically examined some years ago in the Monghyr jail, the Dharhis were found to be the strongest and best nourished. The ostensible means of livelihood of this essentially criminal caste is breeding animals and manual labour, but there is no doubt that in nearly every case the Dharhi is a habitual thief or burglar. It is said, indeed, that they look on thieving as their traditional occupation, so much so that a theft committed by one Dharhi in another Dharhi's preserve, without his consent, is mentioned as a bar on intermarriage. Thirty years ago enquiry showed that out of 1,003 Dharhis in this district no less than 209 had been more than once in jail, while the jail register showed the names of 60 more who could not be identified owing to false names and addresses having been given. The caste now number in this district about 2,200, most of whom are found in the Sheikhpura and Surajgarha thanas.

POLICE.

The marginal table shows the police-stations in the district. The police force consists of the Superintendent, assisted by a Deputy Superintendent, (and sometimes by an Assistant Superintendent), seven Inspectors, two Sergeants, 48 sub-inspectors, 64 Head-constables, and 556 constables. The rural force for the watch and ward of villages, which is maintained from the *chaukidari* tax, consists of 299 dafadars and 3,697 chaukidars. The total strength of the regular police

Subdivision.	Police-stations.
Monghyr ...	Gogri.
	Khagaria.
	Bakhtiyarpur.
	Monghyr Town.
	Monghyr Mufassal.
	Jamalpur.
	Surajgarha.
	Lakhisarai.
	Kharagpur.
	Tarapur.
	Sheikhpura.
	Barbigha.
Begusara ...	Teghra.
	Begusarai.
	Balia.
	Bariarpur.
Jamui ...	Sikandra.
	Jamui.
	Chakai.
	Jhajha.

thus represents one policeman to every six square miles and to every 3,150 persons. Including both dafadars and chaukidars, there is approximately one village policeman to every square mile and to every 534 persons.

JAILS.

There is a juvenile jail at Monghyr, and subsidiary jails at Monghyr, Jamui and Begusarai. The juvenile jail is for male prisoners sentenced to rigorous imprisonment, who at the time of sentence are under twenty years of age. These adolescents receive physical and moral training with a view to their reformation, elementary education in the vernaculars of the province, and technical education in industries which may form a means of livelihood after release. The industries taught are weaving, tailoring, carpentry, tinsmithy, blacksmithy and canework.

The subsidiary jail and the juvenile jail at Monghyr, which are situated in the fort, have accommodation for 306 prisoners. Barracks are provided for 187 and separate cells for 4 juvenile convicts. The hospital accommodates 34 prisoners; and barracks are provided for 17 female convicts, 15 male convicts, 8 civil prisoners, and 41 prisoners under trial. The subsidiary jail at Jamui has accommodation for 37 male and 7 female prisoners, and that at Begusarai for 24 males and 4 females.

CHAPTER XII.

LOCAL SELF-GOVERNMENT.

District and Local Boards.

OUTSIDE municipal areas, local affairs are managed by the district board which has jurisdiction over the whole district and by the local boards which have been constituted for each subdivision. The district board until recently consisted of six official and eighteen non-official members, of whom twelve were elected, with an elected non-official chairman; but it now consists of thirty-seven members, of whom twenty-eight are elected. In the local boards the majority of the members consist of members elected to the district board by the electorate of the area over which the local board has authority, while the remaining members are nominated by the district board. The number of members thus elected and nominated are in the Sadr local board thirteen and four, in Begusarai nine and three, and in Jamui six and two, respectively.

The district board is responsible for the maintenance of roads, bridges and road-side rest-houses and has the general superintendence of primary and middle schools. It is also entrusted with the management of pounds and public ferries, the control over local dispensaries, the provision of a proper water-supply and village sanitation. The work which is being done under these various heads is described in Chapters IV, V and IX of this Gazetteer; and it is unnecessary to describe it again here. To the local boards, which work in subordination to the district board, are delegated the administration of the sums allotted for the construction and repair of village roads, the supervision of local sanitation, and of local cattle pounds and dispensaries.

Income.

In Monghyr, as in other districts, cesses levied under the Cess Act form the principal source of income of the district board; and the incidence of taxation is slightly over four annas per head of the population. The gross rental of the district according to the valuation for cess purposes is Rs. 85,60,000.

The following table shows the income of the board, with the principal sources, for the five years ending with 1923-4.

—	1919-20	1920-21	1921-22	1922-23	1923-24
	Rs.	Rs.	Rs.	Rs.	Rs.
Total income ...	6,33,957	6,63,386	5,95,536	6,08,466	7,57,237
Loans and advances.	34,224	77,172	34,676	44,635	1,43,933
Total receipts ...	6,68,181	7,40,558	6,30,212	6,53,101	9,01,170
Principal heads—					
Local rates ...	4,72,647	4,78,193	4,43,807	4,71,203	5,42,948
Cattle-pounds	16,787	14,195	8,855	12,710	14,116
Ferry tolls ...	49,080	46,208	44,467	44,232	41,721
Contributions from Government.	56,478	84,757	73,966	49,386	1,08,518

The following table shows expenditure for the same period.

—	1919-20	1920-21	1921-22	1922-23	1923-24.
	Rs.	Rs.	Rs.	Rs.	Rs.
Total disbursements.	6,64,008	6,46,559	6,28,987	6,66,738	8,56,901
Principal heads—					
Civil works ...	3,36,771	3,44,334	3,18,233	3,46,961	4,28,991
Education ...	1,01,464	1,34,210	1,30,446	1,28,662	1,32,255
Medical ...	87,566	85,389	84,717	94,419	98,888
Administration	26,099	25,341	26,500	28,119	26,530
Veterinary, etc.	8,714	7,312	10,279	10,454	11,791
Pensions, etc. ...	6,677	7,491	7,612	8,637	9,879
Repayment of loans and advances.	87,928	31,834	36,609	39,360	1,36,999

MUNICIPALITIES.

Monghyr.

The municipality of Monghyr, which was established in 1864, is administered by a municipal board consisting of twenty-five commissioners, of whom twenty are elected, three are nominated, and two are *ex-officio* members. The area within municipal limits is 7·62 square miles, and the number of rate-payers is 9,257, representing 19·7 of the population residing within the municipal area. Important works have been carried out by the municipality in recent years, in the provision of a piped water-supply from the Ganges at a cost of over 5½ lakhs of rupees, and the carrying out of a drainage scheme at a cost of nearly two lakhs. The following tables show the income and expenditure of the municipality for the five years ending with 1923-4.

INCOME.

—	1919-20	1920-21	1921-22	1922-23	1923-24
1	2	3	4	5	6
	Rs.	Rs.	Rs.	Rs.	Rs.
Total receipts	1,45,276	1,54,111	1,26,966	1,73,001	1,53,166
Principal heads—					
Tax on houses...	29,198	32,149	29,761	37,712	34,047
Tax on animals and vehicles.	3,812	3,748	3,467	4,024	3,978
Latrine tax ...	19,017	20,777	19,478	23,863	21,346
Water-rate ...	25,413	28,954	27,857	33,214	29,473
Ferry tolls ...	3,100	3,100	3,100	3,100	3,100
Markets and slaughter-houses.	1,996	2,034	1,504	2,033	1,713
Cattle-pounds	1,358	1,217	710	820	933
Grants from Government.	21,780	23,379	6,992	7,761	14,618
Grants from Local Funds.	11,908	11,700	12,832	10,832	13,082
(Deposits and advances.)	14,435	18,586	12,920	8,493	16,664

EXPENDITURE.*

—	1919-20	1920-21	1921-22	1922-23	1923-24.
1	2	3	4	5	6
	Rs.	Rs.	Rs.	Rs.	Rs.
Total disbursements.	1,54,542	1,65,643	1,37,509	1,52,213	1,83,50[illegible]
Principal heads—					
Administration	8,584	5,912	6,262	7,519	10,023
Lighting ...	2,074	2,324	1,858	1,632	2,849
Water-supply—capital.	...	8,888	...	8,087	21,814
Water-supply—establishment, etc.	22,398	21,071	26,774	28,206	32,338
Drainage—capital.	1,061	10,018	3,868	330	228
Drainage—establishment, etc.	3,952	3,806	3,916	3,736	4,183
Conservancy ...	30,161	28,486	28,996	26,840	29,206
Sanitary officers	5,278	5,463	4,038	1,939	4,980
Hospitals and dispensaries.	31,261	22,266	22,769	31,695	33,940
Markets and slaughter-houses.	727	696	756	1,015	845
Veterinary ...	1,000	1,000	1,000	1,000	1,000
Roads ...	6,328	13,806	6,847	8,564	14,315
Education ...	4,403	5,729	4,679	4,799	4,837
Interest on loans	2,169	2,012	1,848	1,686	1,542
Repayment of loans and advances.	23,253	25,767	17,827	18,565	14,563

The only other municipality in the district is that of Jamalpur, which was constituted in 1883. The municipal board consists of twenty commissioners, of whom sixteen are elected, one is nominated, and three are *ex-officio* members. The area within municipal limits is four square miles, and the number of rate-payers is 3,656, or 14¾ per cent. of the population living within municipal limits. The town is well laid-out and well kept, and the municipality has a high reputation for efficiency. An account of Jamalpur will be found in Chapter XIV below. The following tables show the municipal income and expenditure for the five years ending with 1923-24. Jamalpur.

INCOME.

—	1919-20	1920-21	1921-22	1922-23	1923-24
1	2	3	4	5	6
	Rs.	Rs.	Rs.	Rs.	Rs.
Total receipts...	87,091	54,220	49,992	49,652	48,135
Principal heads—					
Tax on holdings	23,755	23,724	24,706	26,512	29,982
Latrine tax ...	10,323	9,866	11,693	11,721	10,111
Tax on animals and vehicles.	1,372	1,201	1,385	1,215	1,401
Markets and slaughter-houses.	1,698	1,677	1,773	1,803	2,128
Cattle-pounds...	971	643	677	725	844
Grants from Government.	46,185	14,869	2,185	1,185	1,516
Loans and advances.	624	1,048	1,010	184	1,130

EXPENDITURE.

—	1919-20	1920-21	1921-22	1922-23	1923-24
1	2	3	4	5	6
	Rs.	Rs.	Rs.	Rs.	Rs.
Total disbursements.	43,205	67,309	66,611	46,798	51,529
Principal heads—					
Administration	4,902	5,197	6,629	5,195	5,574
Lighting ...	49	42	10	343	411
Water-supply—capital.	...	25,383	15,283	653	...
Water-supply—maintenance.	117	181	184	433	246
Drainage—capital.	2,046	1,298	8,834	2,400	1,289
Drainage—maintenance.	5,185	5,064	5,395	4,522	4,002
Conservancy ...	15,053	14,689	14,915	15,895	18,592
Sanitary officers	1,056	847	1,062	918	1,014
Hospitals and dispensaries.	1,534	1,320	1,664	1,961	2,044
Markets and slaughter-houses.	1,818	1,158	453	512	495
Roads ...	3,304	3,851	4,669	6,637	9,720
Education ...	2,289	2,704	2,701	2,876	4,890
Repayments of advances, etc.	2,367	2,068	2,020	1,715	337

CHAPTER XIII.

EDUCATION.

Progress of Education.

In the year 1874-5, shortly after the introduction of the system of giving grants-in-aid to schools, there were 229 schools in the district attended by 6,675 pupils. By 1881-2 the number of schools had risen to 2,755 and of pupils to 30,403; but there was a falling off in the next decade, the number of schools in 1890-1 being 1,497 and of pupils 25,536. The next ten years witnessed an improvement, the returns for 1899-1900 showing 1,608 schools with an attendance of 32,737; but next year, largely owing to the outbreak of plague, the number of schools decreased to 1,301 and of scholars to 25,738. The table below gives the figures at the end of the last four periods of five years.

Class of institutions.	1909-10		1914-15		1919-20		1924-25	
	Institutions.	Pupils.	Institutions.	Pupils.	Institutions.	Pupils.	Institutions.	Pupils.
1. PUBLIC INSTITUTIONS.								
For males.								
Colleges	1	17	1	25	1	55	1	48
High schools	4	890	4	1,388	6	1,599	5	1,399
Middle English schools ...	7	547	13	1,31[illegible]	23	2,103	24	2,197
Middle Vernacular schools...	8	408	6	473	11	850	13	1,056
Upper Primary schools ...	86	4,516	96	5,[illegible]19	119	6,69[illegible]	156	9,500
Lower Primary schools ...	970	25,463	1,123	28,090	1,261	33,566	1,439	39,553
Special schools ...	13	481	20	392	15	349	19	885
Total	1,089	32,322	1,263	36,79[illegible]	1,436	45,2[illegible]2	1,657	54,638
For females.								
Upper Primary schools ...	3	123	6	320	12	509	11	4[illegible]5
Lower Primary schools ...	32	794	111	2,484	205	4,629	208	4,485
Total	35	917	117	2,804	217	5,138	219	4,950
Total public institutions	1,124	33,239	1,380	39,602	1,653	50,350	1,876	59,588
2. PRIVATE INSTITUTIONS.								
For males	289	3,469	239	3,704	158	2,[illegible]92	111	2,773
For females	...	...	...	...	2	43	1	32
Total	289	3,469	239	3,7[illegible]4	160	3,[illegible]35	112	2,805
Grand Total of all institutions.	1,413	[illegible]6,[illegible]08	1,619	43,3[illegible]6	1,813	53,385	1,988	6[illegible],393

The figures for primary schools include *maktabs* and Sanskrit *patshalas*: in the former case the curriculum is supplemented by religious instruction on the Islamic system and in the latter by the study of elementary Sanskrit.

Literate population.

At the time of the census of 1901, the percentage of males shown as literate was 5·8 and of females 0·2. In 1911 the figures were 7·7 and 0·4 and in 1921 they had improved to 9·4 and 0·6, the test of literacy being ability to write a letter to a friend and to read the reply. Among the 21 districts in the province Monghyr occupies an intermediate position, coming ninth in order of literacy for males and eleventh in order for females. The actual number of literate males in 1921 was 94,749 and of females 5,919: of these 5,436 and 554, respectively, were literate in English.

Colleges.

The Diamond Jubilee College at Monghyr with its 48 students is the only college in the district, and teaches only up to the Intermediate standard in Arts. It was opened in the year 1898 to commemorate the Diamond Jubilee of Queen Victoria, among those who contributed to the cost being Raja Kamleshwari Prashad Singh, Rs. 6,000; Maharaja Bahadur Sir Ravaneswar Prashad Singh of Gidhaur, Rs. 3,000; Raja Ashutosh Nath Roy of Karanchaura, Rs. 3,000; Musammat Shyam Kumari of Ulao, Rs. 3,000; Raja Bahadur Ram Narayan Singh of Khaira, Rs. 2,000; Babu Ganga Prashad, Rs. 1,000; and Maharaja Bahadur Sir Harballabh Narayan Singh of Sonbarsa, Rs. 1,000. Musammat Shyam Kumari of Ulao also created an endowment yielding Rs. 1,200 and Rai Bahadur Lachmi Prashad Singh one yielding Rs. 770 annually. The college was founded by the amalgamation of the three local high schools, the Zila school, the Kamleshwari Prashad Institution and the Victoria Jubilee School, and the necessary buildings for the classes and the hostel were constructed close to the Zila school: In the year 1905, in consequence of the recommendations of the Indian Universities Commission, the school classes were separated and again became a Government Zila school: this affected the finances of the college and at the same time the outbreak of plague affected the roll which had reached a maximum of 66 but fell to 17 in the year 1909-10. In 1915 the college and hostel buildings were purchased by Government for the use of the Zila school and the college moved to the old building of the Victoria Jubilee School. This, however, was patently unsuitable and four years later the college moved out to Pirpahar, where it was given the use of several buildings which had been designed to form part of a central jail; at the same time the college was given a substantial monthly grant by Government. The distance from the town, however, affected the numbers and the tenure of the site being precarious it was decided in the year 1920 to collect funds for new buildings which should commemorate the visit to India of His Royal Highness the Prince of Wales. The subscribers included Raja Shivanandan Prashad Singh, Raja Raghunandan Prashad Singh, and Rai Bahadur Balmiki Prashad Singh, Rs. 15,000 each; Rai Bahadur Dulip Narayan Singh and Rai Bahadur Ramsumram Prashad, Rs. 5,000 each; Rai Bahadur Sir Hari Ram Goenka and Rai Bahadur Shiva Prashad Jhunjhunwala, Rs. 5,000; and Babu Kedarnath Goenka, Rs. 2,500. A new site of about 14 acres was secured in a position close to the town and with the help of a grant of Rs. 21,350 from Government new buildings to be called the Prince of Wales building and the Johnston hostel, costing in all about one lakh of rupees, were completed in the year 1925. These buildings will accommodate some 200 scholars and 36 boarders and it remains to be seen how far their completion will affect the roll number. The college is managed by a Governing Body consisting of the District Magistrate, the District Judge, the Inspector of Schools, the Principal of the college, a representative of the staff, and five other non-officials including Raja Devakinandan Prashad Singh, who is the Secretary.

Secondary Education.

The five high schools are the Zila school at Monghyr, which is managed by Government: the aided schools at Jamalpur, Begusarai and Jamui, and the unaided Town school in Monghyr. Of the 24 middle English schools 22 were until recently aided by Government and two were unaided: the control of these schools, however, is about to be handed over to the district board. Of the 13 middle vernacular schools nine are managed by the district board and the other four receive grants from the same source.

Primary schools for boys.

The primary schools for boys are mainly of the class known as stipendiary, the teacher or *guru* being the manager of the school, choosing its home (which is often a very unsuitable one), charging fees, and receiving a small stipend from the board. Endeavours are being made gradually to replace this rather haphazard system by one of schools under the direct management of the board. It will be noticed that the primary education of boys is making steady progress.

Special schools.

The 19 special schools are the Jamalpur technical school; seven elementary training schools which can train 17 teachers apiece for work in primary schools; two madrasas or advanced schools teaching Persian and Arabic; seven *tols* or advanced schools teaching Sanskrit; a school in the Juvenile Jail and one attached to the Peninsular Tobacco Company's factory. Of these the first only is of any great importance.

The Jamalpur Technical school.

The Jamalpur Technical School was inaugurated about the year 1867 as a night school for European and Anglo-Indian apprentices, the teaching being carried out by the staff of the boys' day school in existence at that period. In 1876 it was brought under Government inspection and a grant-in-aid was sanctioned. A class for about 12 Indian apprentices was added in the year 1920 and the so-called night school was then converted into a day school.

In 1922 a scheme was drawn up by the East Indian Railway to improve the training of the apprentices in the locomotive department and to provide similar training for an additional number of Indian apprentices. The Government of Bihar and Orissa agreed to pay a lakh and a half of rupees towards the non-recurring cost and one-fifth of the recurring expenditure, up to a limit of Rs. 42,000 per annum, for ten years.

The new scheme came into force in October 1924. Provision has been made for the training of 109 Europeans and 101 Indians. Two-fifths of the total number of places for Indians, or 40 such places in all, have been reserved for Indians domiciled in Bihar and Orissa. European students are boarded free of charge and Indians receive an allowance sufficient to cover the cost of living. The course of training is for five years and apprentices are paid monthly stipends during their apprenticeship. For Europeans the rates are Rs. 10 in the first year, Rs. 15 in the second, Rs. 20 in the third, Rs. 25 in the fourth and Rs. 30 in the fifth : for Indians the amount is half as much in each case. All the apprentices are treated on the same footing and there is no distinction of status in their training and work. The extra allowance to European and Anglo-Indian apprentices is paid by the East Indian Railway.

The apprentices are trained primarily with a view to their being employed afterwards on the East Indian Railway but only some of them can be so employed. No guarantee is given; but the railway administration has undertaken to employ them to the greatest possible extent.

The management of the school is in the hands of a committee consisting of eight members, of whom seven are Railway officers and one is a representative of the Government of Bihar and Orissa.

Private institutions. The private schools, which do not conform to the standards or teach the courses prescribed by Government, are of no particular importance and, as the figures show, are being gradually absorbed into the departmental system.

The education of girls. The education of girls is not carried beyond the primary stage anywhere in the district; and the figures, particularly those for the past five years, are disappointing. There were 4,931 girls in girls' schools at the end of the year 1924-5, and 2,712 in boys' schools. The primary schools for girls are mostly of the stipendiary class, like those for boys.

The education of Muhammadans. The number of Muhammadans at school was 3,006 in 1904-5, 5,612 in 1914-5, and 6,756 in 1924-5. The percentage of Muhammadans at school is 3·6, against a percentage of 3·0 for the whole population of the district.

The education of the depressed classes. At the end of the year 1924-5 there were 2,200 aborigines at school, 188 children belonging to the untouchable castes and 117 belonging to criminal tribes. There is one special school in the district intended for the education of boys of the *dom* caste.

European education. There are day schools at Jamalpur and Jhajha for European and Anglo-Indian children, both maintained by the East Indian Railway. At the end of the year 1924-5 the school at Jamalpur had 49 boys and 38 girls on the roll; and the school at Jhajha 13 and 7. The former school cost Rs. 10,706 of which Government paid Rs. 3,627; at Jhajha the cost was Rs. 3,223 and the Government grant Rs. 474. Figures for these schools are not included in the table given at the beginning of this chapter.

Inspecting staff. The district shares with the rest of the Bhagalpur Division an Inspector of Schools, an Assistant Inspectress and a Special Inspecting Officer for Muhammadan Education. Its own staff consists of a District Inspector, two subdivisional deputy inspectors, 13 sub-inspectors and two inspecting *maulavis*. The latter are in charge of the *maktabs*.

Hostels.

At the end of 1924-5 there were 13 hostels in the district and the number of boarders at the end of 1924-5 was 443. These figures include the hostel of the Diamond Jubilee College with its five boarders.

Libraries and literary societies.

The only institutions coming under this head are the European Institute at Jamalpur and the Monghyr Bar Association. Neither is of special importance.

CHAPTER XIV.

GAZETTEER.

Bahadurpur.—A village in the north of the Begusarai subdivision, situated close to the boundary of the Darbhanga district. It contains the residence of a well-known family of this district. According to the family chronicles, its founder was one Lakshmi Sah, who came here from Rajputana and acquired a grant of land extending, it is said, from Manjhaul to Mohraghat. The last proprietor of the estate was Lakshmi Prasad Singh, on whom Government bestowed the title of Rai Bahadur. His son Udit Narain Singh is the present proprietor. The area of the estate is about 53,600 acres, and the rent-roll is about Rs. 1,40,000.

Bakhtiyarpur.—A village in the extreme north of the Monghyr subdivision, with a railway station, called Makhana-bazar, on the Mansi-Bhaptiahi extension of the Bengal and North-Western Railway. It contains a police outpost and district board bungalow, and is the headquarters of an old Muhammadan family known as the Chaudhris of Bakhtiyarpur. The following sketch of its history has been prepared from an account furnished by the family. The founders of the family are said to have been two brothers, Sheikh Golan and Sheikh Achhay Siddiki who came from Jaunpur in 1009 F. (i.e., about 1600 A.D.), and married the daughters of two brothers, Chaudhri Farid and Chaudhri Hafiz, who then held *tappas* Salimabad (*pargana* Ballia), Simri, Kachaut and Hamidpur. Chaudhri Farid, it is said, offered to give them a grant of *tappa* Simri and shares in the other *tappas*, if they cleared the jungle and brought the land under cultivation. This they did, destroying the wild beasts which then infested the country. Subsequently a dispute arose with the neighbouring zamindar of Nisankpur Kurha, and this led them to go to Delhi and obtain a royal charter (*farman-shahi*) from Shah Jahan.

According to the family records, the estate was held by the descendants of the brothers, of whom there is nothing of interest to record until 1126 F., when it passed to Gulam

Muhammad. The latter had some difficulty in retaining possession of it owing to the aggression of one Deo Karan Singh Dundia of Tirhut. Gulam Muhammad brought a complaint before the Nawab of Monghyr, Sarmast Ali Khan, who, pleased with his address, offered him a writer's post (*kalamdan munshigiri*), granted him the sanad of the estate, and himself defeated Deo Karan Singh. Gulam Muhammad appears to have become a member of the court of the Nawab, and when he visited the Emperor, accompanied him to Delhi, making over charge of the estate to his brother Gulam Ali, who in his turn handed it over to his brother Hedayat Ullah. The latter had to meet the same kind of trouble as Gulam Muhammad, being attacked by one Rup Narayan Singh, who tried to seize *pargana* Pharkiya. Hedayat Ali succeeded in ousting him, but was soon afterwards treacherously killed. Rup Narayan Singh then looted his residence and destroyed the old *sanads* and *farmans* granted to his family. Gulam Ali subsequently recovered the estate under the orders of Ali Vardi Khan. In 1194 F. (1791 A.D.) Chaudhri Muhammad Ali moved the family residence from Simri to Bakhtiyarpur, where his descendants still reside. The estate is now held by Chaudhri Muhammad Nazirul Hassan, and the rent-roll is Rs. 82,000.

Bamdah.—A village in Chakai thana, four miles north of Chakai. It contains a station of the United Free Church of Scotland Mission to the Santals, the site of which was chosen by the late Dr. James Dyer of that mission, when he visited the place from Pachamba in 1878. A lease of seven acres was obtained, and in 1880 two preachers took up their residence in some mud huts, the construction of a bungalow being commenced in 1884. Since that year the work of the mission has developed considerably. The Rev. J. M. Macphail, M.D., began to reside at Bamdah in 1890; in 1894 a hospital was built, and by the end of 1900 a church with a handsome belfry was erected. The enlargement of the hospital was undertaken as soon as the church was completed, when an operating room was provided, constructed on modern principles. At the same time additional ward accommodation was provided by building a new dispensary and consulting room; and more recently the hospital has been enlarged and a small bungalow has been added for the accommodation of private patients. Three thousand four hundred operations were performed at the hospital in 1923, of which 2,890 were eye operations, including 1,426 for cataract. Mrs. Macphail, L.R.C.P., L.R.C.S. (Edin.), has assisted in the work at Bamdah since 1898; and a son, Mr. Ronald M. Macphail, M.B., M.CH., was appointed an additional member of the mission in 1924.

The educational work of the mission is also important. There are 64 schools connected with the mission, of which 62 are village schools, more than half of them being night schools. Six hundred and sixty-four boys and 294 girls were enrolled in 1923.*

Banaili Raj.—An extensive estate in the districts of Monghyr, Bhagalpur and the Santal Parganas. The founder of the family which owns this estate was Hazari Chaudhri, a *tahsildar* of the Passara Raja of Purnea, who, about 1780 A.D., acquired by purchase *pargana* Tirakhurda in that district. His son, Dular Singh, acquired property in Monghyr, Bhagalpur and Malda about the year 1800 A.D., and on his death, after some litigation, the estates were divided equally between his two sons, Raja Bidyanand Singh and Kumar Rudranand Singh, grandfather of the Srinagar Kumars. Raja Bidyanand Singh subsequently purchased Mahalat Kharagpur, and in 1851 was succeeded by his son, Raja Lilanand Singh Bahadur, who also added to the estate by purchasing Chandpur Husain and *taluk* Khajuria. Raja Lilanand Singh died in 1883, and was succeeded by his son, Raja Padmanand Singh Bahadur. He also left a minor son, Kumar Kalanand Singh and a posthumous son, Kumar Kirtyanand Singh, on whose behalf a suit was instituted in 1888. It ended in a compromise decree, under which Kumar Kalanand Singh and Kumar Kirtyanand Singh were to be owners of 9 annas of the Banaili Raj, while Raja Padmanand Singh kept the remaining 7 annas. In 1903 Kumar Chandranand Singh, son of Raja Padmanand Singh Bahadur, brought a suit against his father for partition and other reliefs, which also ended in a compromise decree, under which the son was declared to be owner of a 3½ annas share and Raja Padmanand Singh owner of the remaining 3½ annas share. Subsequently, however, in September 1905, Raja Padmanand Singh Bahadur transferred his rights and interests to his son Kumar Chandranand Singh, who died in the year 1908, leaving a widow Srimati Chandrabati *alias* Gangabati. After she succeeded

---

*** An account of the Mission will be found in *Santalia*, by Dr. Macphail who is also author of *The Story of the Santal* (Thacker, Spink and Co, Calcutta).**

to the 7 annas share, she was declared a disqualified proprietress on her own application; and the Court of Wards is in possession of the 7 annas share on her behalf. The Court of Wards has again given lease of the 7 annas estate for twelve years from 1324 F.S. to the proprietors of the 9 annas share, who are the principal creditors of the 7 annas estate.

In 1919 the title of Raja Bahadur was given to Kumar Kirtyanand Singh, who has been a member of the Provincial Legislative Council for the last sixteen years. Kalanand Singh, who also had the title of Raja Bahadur, died in 1922; and was succeeded by his two sons, Kumar Ramanand Singh and Kumar Krishnanand Singh. The two Kumars and Raja Kirtyanand Singh Bahadur are now in possession of the whole estate, which is managed jointly.

The bulk of the Banaili Raj property in this district forms part of Mahalat Kharagpur, an extensive estate on the revenue roll of Bhagalpur bearing *tauzi* number 445. It consists of the following *parganas* :—(1) Sahroi, (2) Lakhanpur, (3) Kherhi, (4) Sakharabadi, (5) Parbatpara (including *tappas* Lodhwa, Simraon, Dighi, etc.), (6) Wasila, (7) Godda, (8) Hazar Takhi, (9) Amlo-Motia, (10) Handwe (including a *ghatwali* held by the Rani of Handwe), (11) Chandan Katoria, (12) Jahangira and Masdi, (13) Dharahra, (14) Abhaipur, (15) Singhaul and (16) Danda Sakhwara. Mahalat Kharagpur, together with *parganas* Kajra and Haveli Kharagpur and the Lakshmipur estate in the Santal Parganas, originally formed part of the territory held by the Raja of Kharagpur, of whose family history an account will be found in the article on Kharagpur. When Raja Rahmat Ali Khan fell into arrears of revenue, in 1840 Mahalat Kharagpur, with the *ghatwali* of Handwe, was sold and purchased by Raja Bidyanand Singh, grandfather of the present proprietors of the Banaili Raj, and by Balnath Sahu of Bhagalpur, who in 1841 transferred his share to Raja Bidyanand Singh Bahadur. Since then there have been constant disputes and litigation in connection with this *mahal*, especially with the Darbhanga Raj, which in 1848 purchased at a sale for arrears of Government revenue Haveli Kharagpur, which is surrounded by Mahalat Kharagpur. There has also been much litigation concerning the *ghatwali* tenures, the results of which have been mentioned in Chapter X. Most of those tenures have now been converted into *mukarari istamrari* tenures, and only a few, including the *ghatwali* of Handwe in the Santal Parganas and Kakwara in the Banka subdivision of Bhagalpur, retain their *ghatwali* character.

Begusarai.—Headquarters of the subdivision of the same name situated in 25° 26′N. and 86° 9′E. about five miles to the north of the Ganges. The town is situated at a distance of about half a mile from the railway station, and the Tirhut road runs east and west through it. This is an old road shaded by an avenue of interlacing trees, which in the main follows the course of the railway; but it has fallen into disuse to a great extent since the latter was constructed. Approaching Begusarai from the east along this road, the first building met with is a dak bungalow with four rooms, erected on a low-lying piece of land at a distance of about half a mile from the town itself. A tablet let into a pillar of the verandah shows that the flood of 1904 rose to a height of nearly three feet above the plinth. Half a mile beyond the bungalow, on the outskirts of the town, lies the subdivisional office, a small building with an upper storey. The building is situated in a fine compound, in which mango trees grow in profusion; and the Munsif's court, schools, hospital, thana, and local board offices are situated close by. Just-beyond the Sub-divisional Officer's compound to the north is a fine tank fringed by palm trees, at the south-west corner of which is a small temple, where Kayasths worship.

The town contains two mosques and three temples, but none of these have any features of architectural interest, and the bazar itself is much the same as other Bihar bazars. Half-way through the latter is the *sarai*, from which the town presumably derives its name. It consists of an open space surrounded by tumble-down shops and lodging-houses. Beyond the *sarai* the town opens out, the avenue of trees is seen again, and presently on the western skirts of the town we get a glimpse of a pleasant bungalow, seen through an avenue of bamboos.* This is the Begusarai Factory, better known locally as Harrakh, which was built by Mr. James Hennessy in 1863, and is now the property of Babu Khub Lal Singh. There is a thriving grain market with several godowns at the southern extremity of the town, and to the east and south-east are extensive mango groves.

* The above account was contributed by the late Mr. H. R. T. S. Perrott, I.C.S., formerly Subdivisional Officer of Begusarai.

The population of the town, according to the census of 1921, is 9,062. It was made the headquarters of the subdivision by a notification dated the 6th January 1870.

Begusarai Subdivision.—The north-western subdivision of the district lying between 25° 15′ and 25° 47′N. and between 87° 47′ and 86° 27′E. It has an area of 751 square miles, and is nearly 34 miles long from north-west to south-east and 22 miles broad from north to south. It is bounded on the north by the district of Darbhanga, on the east by the northern portion of the Monghyr subdivision, on the south by the Ganges, and on the west by the Samastipur subdivision of the Darbhanga district. The subdivision is a flat alluvial plain, almost entirely under cultivation. It is intersected by a number of rivers, of which the principal are the Burh Gandak, Balan, and Bairanti, all subject to floods in the rains. Along the northern bank of the Ganges there are two embankments, one maintained by the Public Works Department, and another, called the Gupta Bandh, remodelled by Government, which protect the adjoining country from the floods of that river. There are also small embankments on the banks of the Burh Gandak constructed and maintained by zamindars; but in years of heavy floods these embankments are breached and the low lands submerged. To the north there is a large swampy lake, called the Kabar Tal, about eight miles in length and two miles in breadth, which contains deep water throughout the year.

The subdivision is served by two lines of the Bengal and North-Western Railway, one, the line from Simaria Ghat to Muzaffarpur and Darbhanga, passing through the western portion of the subdivision, and the other, the Katihar line, extending from Barauni junction on the west to the Sahebpur Kamal station on the east. The latter line passes through the south of the subdivision, and the subdivisional headquarters are situated on it. The population was 586,199 in 1921. There are altogether 755 villages, one of which, Begusarai, is the headquarters.

Bhaduria-Bhur.—A hot spring about two miles east-south-east of Rishikund, on the other side of the Kharagpur range of hills, not far from the village of Dariyapur in the Jamalpur thana. The name means the cleft of Bhaduria Hill, and the spring is much cooler than the Rishikund spring, of which it is locally believed to be a branch. It emerges at the foot of the Bhaduria Hill from among masses of quartzite rocks, accompanied by a free discharge of gaseous bubbles, devoid of smell and uninflammable. The water is drunk by men and cattle. Its temperature as recorded by Colonel Waddell in March 1890 was 98·5°F.

Bhimbandh.—A village in the south of the Monghyr subdivision, situated about 12 miles south-west of Kharagpur and four miles north of Guddih. Close to the village are some hot springs, called Tatal-pani, which are by far the finest in the district.

The following account of these springs is given by Captain Sherwill :—" The first spring is situated about 300 yards to the north of the village immediately under a small detached hill named Mahadeva, from whose base the water issued in a fine stream at a temperature of 147° Fahrenheit. A few hundred yards farther to the north, at the foot of the hornstone hill Damdama, we came upon a region of hot springs. Hot water appeared to be spouting from the ground in every direction; the principal springs, of which there are eight or ten, had uniform temperature of 145°, all rising within a space of about 300 yards square. Across numerous hot streams are, of course, many foot-paths used by the cultivators round about Bhimbandh, but nowhere at the point of crossing did I find the water above 120°, and even that temperature made the men and women hurry across the stream when fording from bank to bank. To our European skins the heat of 120° was intolerable, nor could any of the party walk coolly across any of the fords at that temperature without being severely scalded though not blistered. Luxuriant crops of rice are raised by the aid of the hot streams, large fields being fed by the water, but at a reduced temperature by leading it in devious courses to the cultivated land. The united waters of all these hot springs are conveyed away by the small stream called the Man ".* Not far off is a small pool of cold water under an overhanging rock in that river, called Bhimkund, which is sacred to Bhim and is visited by pilgrims.

These springs, rising at about 300 feet above sea-level are the principal source of the Man itself. The highest temperature recorded by Dr. Buchanan on the 21st of March, 1811, was 150°. Sherwill in September of 1847, Waddell in

* Capt. S. R. Sherwill, *The Kurrukpoor Hills*, J. A. S. B., Vol. XXI, 1852.

January of 1890, and Schulten in August of 1913, observed temperatures of 147°, 146·2°, and 148°, respectively; but Mr. V. H. Jackson considers that there are twelve sources in the Mahadeva group and at least nineteen in the Damdama group; and the hottest of them may not have been observed. Readings taken between 1912 and 1919 varied from 145·5° to 146° in the Mahadeva, and from 148° to 148·8° in the Damdama series.

Buchanan noticed that water of the Man, near the springs, was warmer than the atmosphere; and in one place where bubbles were rising in the stream his thermometer registered 98°. Mr. Jackson has traced this to a second series of hot springs along the course of the river, commencing immediately below the Bhimkund and extending for more than a quarter of a mile before the outflow of the first series is reached. Their position varies to some extent from year to year after the rains; but when they can be observed above the stream level their temperature is fairly uniform, though not higher than 133°.

General Cunningham identified the Mahadeva hill with one mentioned by Hiuen Tsiang in the seventh century A.D. as the site where Buddha overcame the Yaksha Vakula. Hiuen Tsiang describes the place as a small solitary double-peaked hill, or, according to another translation, a hill "with successive crags heaped up", situate on the western frontier of Hiranya Parvata, a tract held by recognized authorities to coincide approximately with the hilly portion of this district. To the west were six or seven hot springs, the water of which was extremely hot. Colonel Waddell has shown, however, that there are good grounds for doubting this identification, and that the natural features of the country do not agree with the description of the Chinese pilgrim. He points out that the hill is not on the western but the southern frontier of Hiranya Parvata; and that the hot springs are not to the west of the hill, but actually upon the hill itself and on its eastern and north-eastern slope. There are no remains of any kind except those of a small brick shrine about four feet square housing a linga; there is no history of there ever having been any remains; and the situation is so remote that had they ever existed, it is scarcely possible that every trace of them would have been swept away.*

* L. A. Waddell, *Discovery of Buddhist Remains at Mount Uren in Mungir (Monghyr) District*, J. A. S. B., Vol. LXI, Part I, 1892.

Brindaban.—*See* Kiul.

Chakai.—A village in the extreme south of the Jamui sub-division, situated 31 miles south-east of Jamui. The village contains a police-station, district board bungalow, and a dispensary. The dispensary was built by the district board, which in 1895 agreed to erect it and give a monthly grant for drugs, if the missionary at Bamdah would visit it and dispense medicines once a week.*

A reference to the survey map will show an entry of a fort called Hastings Fort against Chakai. Little is known about its history, but Mr. F. Bradley-Birt, I.C.S., has kindly contributed the following note on the subject:—"It is not clear how this fort came to be known as Hastings Fort. The villagers do not know it by that name, but as the Sarkari Garh or Fatehgarh—the Fort of Victory. It has been generally supposed that the name commemorates the Marquis of Hastings. There is nothing, however, to associate him in any way with this fort, and it must undoubtedly have been in existence considerably before his time. It was apparently built by Captain Browne, the first representative of British rule in this corner of the district, whose name is still remembered by all the people round. I found an old man there whose grandfather was one of the sepoys with Captain Browne at the time the fort was built. It is quite probable that the name of Warren Hastings, the greatest Englishman of the day in India, should have been given to the fort, and there being no evidence, so far as I can discover, to connect it with the Marquis of Hastings, I think this may be taken as one of the few places outside Calcutta that still perpetuates the name of the first great Governor-General. Of the fort itself nothing remains but the bare outline, which can still be traced in mound and trench. Close by is another fort, supposed to have been the native one, of yet vaguer outline, with an open space still known as the Chandmari (rifle range) beside it."

Captain Browne held charge of the Jungleterry (Jungle Tarai) district, in which this tract was then included, from 1774 to 1779, his duty being to introduce peace and settled government. According to local tradition, he was opposed by the *ghatwals*, and their opposition led to the construction of the fort at Chakai. The villagers assert positively that the

*** But see page 87 above.**

fort facing the inspection bungalow is the native fort, and that the ruins near the thana and north of the bungalow are all that remain of the British fort. Local tradition also says that he was buried at Chakai, but his tomb, if it exists, cannot be distinguished from the Muhammadan tombs scattered about.

Chandisthan.—*See* Monghyr.

Chautham.—A village in the north-east of the Monghyr subdivision, situated at the junction of the Baghmati and Tiljuga rivers, about 15 miles north-east of the town of Monghyr. It contains the residence of an old Kshattriya family known as the Babus of Chautham. According to the family tradition, they are descended from Murar Sahi, who, with his brother Jorawar Sahi, was in the service of one Raja Kanchan. The estate of the latter having been taken by the Mughal Emperor Akbar, the two brothers returned to their home near Delhi and rendered good service to the Emperor. They were rewarded by the grant of two zamindaris, Godhna-Manjhi (also called Chirand Chapra) in the Saran district, and *tappa* Chautham in this district. Jorawar Sahi took possession of the former and Murar Sahi of the latter, which he enjoyed till 986 F., i.e., 1578 A.D. He then lost his life in a battle against one Saiyad Sahi, and all the females of his family, for fear of ill-treatment, set fire to the house and were burnt to death. One son, Ram Sahi, however, escaped, and going to Delhi, had the zamindari restored to him. The estate, it is said, was originally covered with jungle, which the founders of the family cleared.

The *tappa*, which comprises 42 *mauzas*, has now been divided among their descendants, and the annual rental of all the shares is said to be about Rs. 1,50,000. The leading representative of the family at present is Babu Keshwar Narayan Singh, who is 11th in descent from Murar Sahi.

Dakra Nullah.—*See* Monghyr.

Deoghar.—A small hill in the south-east corner of the Monghyr subdivision, situated about ten miles south of Kharagpur close to the Sangrampur road. It is composed of giant boulders piled one on top of another, and on its summit is a temple known as the temple of Ochnath, which is dedicated to Siva. The people of the neighbouring villages come here on the 14th day of Phalgun (February) to worship the god, and on this occasion there is a fair lasting for three days. Buchanan Hamilton 100 years ago described the hill as a very picturesque rock of granite. "On its summit is a small temple, to which none of my Hindus would ascend, although they were very desirous, and although a Moslem lascar showed them the example ; but the precipice is tremendous, and the ladders were very bad."

Dilawarpur.—*See* Monghyr.

Gidhaur.—A village and police-station in the Jamui subdivision, situated nine miles south-east of Jamui and one mile from the railway station of the same name. Population (1921) 1,859. It contains the residence of one of the oldest of the noble families of Bihar, the founder of which was Bir Bikram Sah of the Chandel sept of Chandrabansi Rajputs. The earliest account of his ancestors represents them as holding a small estate called Mohaba in Bandelkhand, and being overcome by Prithwiraj, the last Hindu emperor of Delhi. Subsequently, they were driven out by the Muhammadans in the 11th century, and won for themselves three principalities, viz., Bijaigarh, Agorhi-Barhar in the Mirzapur district, and Bardi in the present Rewah State. A few generations later Bir Bikram Sah, a younger brother of the chief of Bardi, left his home under the direction, it is said, of the god Siva, and set out with a large following to the shrine of Baidyanath at Deoghar. Bir Bikram Sah found the country in the possession of aboriginals, said to have been Dosadhs. Them he overcame, killing their chief Nagoria ; and the pool in which the Rajput victor washed his bloody sword is known to this day as the Khandwa Pokhar, i.e., the pool of the sword. He appears to have been the first Rajput invader of this part of Jharkhand, or the forest land, and after the lapse of six centuries the family which he founded is still wealthy and influential.

Their original home was at the foot of the hills, where the remains of an old stone fort and other buildings may still be traced in the scrub jungle ; and close by are the remains of a large fort called Naulakhagarh, described later in this chapter. In time their territory extended further west into more fertile country. The son of Bir Bikram Singh, Sukhdeo Singh, is said to have built at Kakeswar, four miles east of Gidhaur, 108 temples to Siva and one dedicated to Durga. The eighth in descent, Puran Mal, established himself at

Lachhuar (q.v.), sixteen miles west of Gidhaur, and built in 1596 the great temple of Baidyanath, a Sanskrit inscription in which refers to him as *nripati*, or king of men. After his death the property was divided between his two sons, the partition being explained by a quaint legend. It is said that a bard of Delhi came to Puran Mal, and recited some wonderful verses in praise of the Raja. When the time came to reward the bard, he declined to take any remuneration except the "philosopher's stone" (*paras*), which has the power of turning iron into gold. Puran Mal had no such stone, but one day, when he happened to be turning up some earth with a knife, the knife at once turned into gold. The Raja at once saw that the earth must contain the *paras*, and digging it up, gave it to the bard, who took it to Delhi. There he proclaimed his good fortune, and was sent for by the Emperor. The bard said that he would show the stone only in a boat, and it was arranged that the Emperor should sit in one boat and the poet in another. The latter then asked the Emperor to stretch out his sword, and, having touched it with the stone, threw the stone into the river. The sword was turned into gold, and the Emperor was convinced of the truth of the story. As the stone was lost, the name of the donor of the stone was ascertained, and he was directed to come to Delhi.

Puran Mal had died in the meantime, and his two sons Hari Singh and Bisambhar Singh ruled in his stead. The former was taken to Delhi and, being unable to produce another such stone, was sent to prison. While Hari Singh was in jail, Bisambhar Singh succeeded to the *gadi* and became sole master of the estate. Hari Singh, after some time, pleased the Emperor by his skill in archery and was given the *pargana* of Bisthazari. When he came, however, to take possession, he found Bisambhar Singh installed. An amicable settlement was therefore made between them. Hari Singh obtaining 9 annas of Gidhaur *pargana* and 7 annas of Bisthazari, while the rest was retained by Bisambhar Singh. The former was the ancestor of the Gidhaur Maharaja, and the latter of the Kumar of Khaira. Another and more reasonable account is that Hari Singh was kept at Delhi as a hostage for the good conduct of Puran Mal, but otherwise the story of his return and of the amicable settlement with his brother is the same.

Dulan Singh, the fourteenth Raja, received high honours from the Musalman Government, and the title of Raja was confirmed by a *farman* of the Emperor Shah Jahan, which still exists, bearing date the 21st Rajab 1068 A.H., corresponding to A.D. 1651. In the struggle between the sons of Shah Jahan he took up the cause of Dara Shekoh, and the family still possesses a letter from the latter thanking the Raja for his help, besides a letter from Prince Shuja asking for his aid. After the English assumed the government of Bengal and Bihar, Raja Gopal Singh, the nineteenth in descent, was for a time deprived of his estates, but afterwards recovered possession. The title of Raja was granted by the British Government in 1856 to his descendant Jaimangal Singh in recognition of his services during the Santal rebellion of 1855 ; and in 1861 he was granted an estate yielding a rental of Rs. 3,000 per annum to be held rent-free during his life-time, in acknowledgment of the services he rendered during the Mutiny of 1857. Subsequently, he was granted the title of Maharaja in 1865 and was made a K.C.S.I. in 1866 ; while the title of Maharaja Bahadur was made hereditary in the family in 1877 on the occasion of the assumption by Her Majesty Queen Victoria of the title of Empress of India. He was succeeded by Maharaja Siva Prasad Singh, and the present head of the house is Maharaja Chandra Mauleshvar Prashad Singh Bahadur.

The Government revenue paid by the estate is reported to be about Rs. 20,000, and the income about Rs. 2,50,000. After the accession of the late Maharaja Bahadur Sir Rameshwar Prasad Singh much was done to improve the village of Gidhaur, and it now contains a brick-built bazar, a good school, and a charitable dispensary. The old Srivilas palace has been modernized, and a new palace called the Suknivas has been built. A clock tower, built to commemorate a visit of the Viceroy, Lord Minto, was opened in February 1909.

Gogri.—A village in the Monghyr subdivision situated on the north bank of the Ganges, about 15 miles north-east of Monghyr. It is a station on the steamer route from Patna to Goalundo, and is connected by a road, four miles long, with the Maheskund railway station on the Hajipur-Katihar extension of the Bengal and North-Western Railway. The village contains a police-station, Middle English school, dispensary, post-office and inspection bungalow, besides the estate offices of some influential zamindars. It is an important trade centre, being the main outlet for *pargana* Pharkiya. Grain is the chief article of export, but fish are also exported in fairly large quantities. The village was formerly situated some miles to the west, but the old site was washed away by the Ganges some years ago, when a flood of that river carried away the southern portion of the embankment which protects the place. This embankment has recently been raised and strengthened by the district board, and it has now been brought under the Embankment Act after it has been remodelled

Hasanpur.—A hill in the Monghyr subdivision, situated close to the village of Jaynagar a little to the south-east of Lakhisarai. Tradition states that there was a fort on this hill built by Indradyumna, the last Hindu king of Magadha, and that he buried his treasure here. Remains of buildings constructed of brick and stone have been found here, and in the vicinity there are the foundations of other old buildings, cut stones and idols. A fuller description of the hill and its neighbourhood will be found in the article on Jaynagar. It is possible that this is the small solitary hill with a high double peak or with crags piled one above another, which Hiuen Tsiang mentions as the place where Buddha overcame the demon Vakula.

Husainabad.—A village in the south-west of the Monghyr subdivision, situated three miles south of Sheikhpura. It contains the residence of one of the oldest Muhammadan families of the district, which traces back its descent to Makhdum Sheikh Shams-ud-din, surnamed Faryad Ras Kurraish-ul-Hashmi, a Saiyid of the family of the Prophet Muhammad. According to the family tradition, he was born in Turkey, but having come to India with his uncle Kazi Rafi-ud-din, who held the office of kazi under the Sultan of Turkey, settled in Oudh and died there in 790 A.H., i.e., about 1373 A.D. His son Sheikh Manjhan Shahid was murdered while on a journey to Bihar, and thereupon his widow, with her two sons Sheikh Mustapha and Sheikh Junaid, settled at Sheikhpura. After the death of Sheikh Mustapha, several of the family migrated to Patna, where their descendants still reside; but about 1836 A.D. two of the family, Muhammad

Yahya Khan and his son-in-law Fida Ali Khan left Patna and returned to this district, where they settled at Husainabad. The present representatives of the family are called Nawabs by courtesy, this title having been enjoyed by their ancestors, some of whom were distinguished personages. Two of them, Nawab Shab Khan and Nawab Fida Khan, who met their death during the Maratha raids, were, it is said, Wazirs of the Emperors of Delhi; and it is claimed that the office of Wazir was hereditary in the house. The family was granted an *altamgha jagir* by the Emperor Aurangzeb, and Shah Alam gave Nawab Ali Ibrahim Khan a robe of honour, the appointment of Shash Hazari and the title of Amin-ud-daula Aziz-ul-Mulk.

Ali Ibrahim Khan, the most distinguished member of the family, is frequently and prominently referred to in the *Sair-ul-Mutakharin*, where he is described as "the illustrious and high-born Khan, the beneficent, munificent, excellent, learned, sincere and valiant ". He appears first to have won the favour of Ali Vardi Khan, who invited him to Murshidabad with other persons, distinguished for their birth, rank, learning and talents, and granted him a large pension. Subsequently, he became a trusted courtier of the Nawab Mir Kasim Ali Khan, to whom he proved a good counsellor. He tried to dissuade him against his disastrous expedition into Nepal and also to prevent his engaging in war with the British. It was he who advised against the Nawab stopping the boat laden with arms on its way to Patna which formed a *casus belli;* and later, after the defeat at Udhua Nullah, in vain urged him to release his English prisoners or at least send the women under escort to the British army under Major Adams. After the defeat of Mir Kasim at Patna, Ali Ibrahim Khan remained faithful to his master, and there is an amusing tale of his diplomatic conduct when a quarrel took place between the Nawab and his ally Shuja-ud-daula, the Nawab Vizier of Oudh. Mir Kasim Ali, apparently to bring shame on the latter, assumed the dress of a *fakir*, and Ali Ibrahim Khan, being asked to persuade him to assume his proper dress, appeared before him wearing only a shirt and a pointed cap in place of his usual turban. After this, when Shuja-ud-daula imprisoned Mir Kasim Ali, and his followers deserted Mir Kasim, Ali Ibrahim Khan alone remained loyal, showing then, as on other occasions, a fidelity uncommon in those troubled times. It is said that when asked why he still clung to Mir Kasim in spite of the way in which the latter maligned him, his manly and dignified answer brought tears to the eyes even of the mean-spirited Nawab Vizier.

After the battle of Buxar, when Mir Kasim fled northwards, Ali Ibrahim Khan retired to Murshidabad, and thenceforward was largely involved in palace intrigues. He was appointed *Diwan* to the Nawab Mubarak-ud-daula, and subsequently espoused the cause of Muhammad Reza Khan and effected his release from prison. He fell out, however, with the latter, and was ruined by a palace intrigue. Subsequently, he was offered high offices by the Nawab, Muni Begam and the Governor-General, but declined them all. Later, however, we find that he accompanied Warren Hastings when he went to Benares in 1781; and that after the repression of the rising of Chait Singh, Warren Hastings granted him a *khilat*, confirmed the title of Amin-ud-daula Aziz-ul-Mulk, which had been granted to him by Shah Alam, and made him Judge of Benares. Further details of his life will be found in the *Sair-ul-Mutakharin*, which gives an amusing account of his character and ways, of his knack of making extempore verses, of the taste and elegance with which he wore his clothes and turban, of the amenity of his manners, and of his successes as a lover.

The brother of this nobleman was Ali Kasim, of whom there is little to record, except that, when in 1781 Warren Hastings made his well-known visit to Benares with Ali Ibrahim Khan to bring Chait Singh to reason, Ali Kasim Khan met them at Monghyr, entertained them on a lavish scale, and accompanied them to Patna. His son, Muhammad Yahya Khan, moved from Patna to Husainabad, and the next head of the family was his nephew and son-in-law Fida Ali Khan. The latter earned the thanks of Government for furnishing information regarding the movement of rebels in the Mutiny of 1857, while his son Nawab Ali Khan was made a Khan Bahadur, and granted a certificate in recognition of his conduct as an enlightened and loyal zamindar and of the service he rendered in the famine of 1874.

Indpe.—A village in the Jamui subdivision, situated four miles south of Jamui and two miles north of Khaira. It contains the remains of a large fort attributed by local tradition to Indradyumna (called locally Indradaun), who is believed to have been the last of the Pala kings. The following account of the remains, as they were a century ago, is given by Dr. Buchanan: "The work is pretty extensive, the fort being a square of about 1,650 feet. The rampart of brick has been about 10 feet thick, and the ditch about 15 feet wide, so that neither could have been intended for any serious resistance to an army; but they were sufficient to guard against surprise or insurrection. The east face is rather irregular, being bent in south from the gate, which is not exactly in the middle, as is also the case with the western gate. In the northern and southern faces are no gates. Before the eastern gate are two heaps of brick, that have been considerable buildings. Within the outer fort has been a citadel. To the left of the passage between the outer gate and that of the citadel, entering from the east, are two considerable heaps of brick: that nearest is said to have been a temple of Siva, and a Priapus still remains. On the right, towards the north-east corner of the outer fort, are three very considerable heaps, surrounding four smaller. Towards the south-west corner of the inner fort, on its south side, is another heap; and these are the only traces of buildings in the outer fort.

"On entering the citadel from the east, you have on the left a mound, which from its great height is by far the most conspicuous part of the whole building. It is said to have been a place (*chandini*) to which the Raja repaired to enjoy the freshness of the evening air; and this tradition is confirmed by the remains of a small terrace of brick, as usual in such places, that has been built on the top of the mound. The mound is, however, so very great a member of the whole, that I rather suspect it to have been a solid temple of a Buddh: as we know that the Rajas of this part of the country, immediately previous to the Muhammadan invasion, were of that sect. Beyond the mound is the royal palace, as it is called, raised on a lofty terrace 220 feet long by 110 wide. Traces remain to show that this terrace has been occupied by three apartments, where probably the Raja sat in state, while his family was lodged in wooden buildings, that have left no trace. The brick buildings in the outer fort, and without the eastern gate, were probably public offices, and the officers and domestics of the family were perhaps accommodated in buildings of no durability sufficient to leave traces that are now observable."

The great mound referred to in this account marks the remains of an old stupa measuring 125 feet in diameter at its base by 35 feet in height. The base is now broader than it was owing to the accumulation of debris, but the diameter of the mound is 65 feet at a height of only 20 feet from the base. It stands within an irregular enclosure of two walls and has a deep shaft sunk from the top. The place was clearly an extensive settlement, and many mounds still remain which would repay exploration. In fact, Mr. Beglar, who visited the place in 1872, says that he can "confidently assert that a careful and thorough examination of the mounds, and especially of the great tope, will prove a great acquisition to the present meagre knowledge of ancient Indian structures ".

Jaimangalgarh.—*See* Kabar Tal.

Jamalpur.—A town in the Monghyr subdivision, situated six miles south of Monghyr and 299 miles from Calcutta, at a height of 173 feet above sea-level. The town, which is picturesquely situated at the foot of the Kharagpur Hills, owes its development to its being the headquarters of the Mechanical Department of the East Indian Railway. The workshops were first established here in 1862. The place was at one time the headquarters of the Engineering, Traffic, and Locomotive Running Departments; but these Departments have been removed to Calcutta. The workshops are among the largest in India, covering an area of 99 acres, of which 26 acres are roofed over. The value of the outturn is about 1½ crores of rupees a year; and employment is given to about 278 Europeans and 11,500 Indians. The iron foundries are capable of a monthly outturn of over three thousand tons of castings, while the steel foundry can turn out eight hundred tons of steel, and the rolling mill a thousand tons of rolled iron and steel bars monthly. The machinery is principally driven by electricity, and the same generating plant supplies current for electric lights and electric fans in the staff quarters and for the Railway pumping plant at Monghyr .

The Indian workmen are chiefly recruited from Jamalpur, Monghyr, Dharhara, Abhaipur, Kajra, Bariarpur, Sultanganj and adjoining villages. Those who live outside Jamalpur are conveyed daily to and from work by special workmen's trains.

The bazar, which is separated from the European quarter by the railway, presents no features of interest. The European quarter consists almost entirely of bungalows owned by the Railway, and is neatly and prettily laid out. The station is provided with Protestant and Roman Catholic churches, a Union church and Institutes, containing libraries, reading and billiard rooms and entertainment halls. There are also tennis courts, a swimming bath, a bowling green, and golf links. A technical school with hostels for apprentices, and four schools for the education of the children of employés are provided; and there is a hospital and dispensary. There is no hotel or dak bungalow in the town, the only accommodation for travellers being at Messrs. Kellner and Company's retiring rooms at the railway station.

Jamui.—Headquarters of the subdivision of the same name, situated four miles south-west of the Jamui station on the main line of the East Indian Railway. Population (1921) 3,977. The town is connected by a good metalled road with the Jamui railway station, which is situated in Mallepur village; and it consists for the most part of one long street, forming a continuation of the Mallepur road. On either side lie the shops and dwellings of the grain merchants and other shopkeepers, while minor roads branch out at right angles, leading to the non-trading parts of the town and to the residences of Indians of the professional classes.

On approaching Jamui from the railway station, the first public buildings met with are the high English school, the Subdivisional office and the Magistrate's residence—masonry buildings of the ordinary kind. On the opposite side of the road is the Munsif's Court. About a quarter of a mile within the town, in a clear open space, stands the police-station. Immediately behind it is the sub-jail, and behind this again the charitable dispensary, which was completed in 1874. Although it lies within the great Gangetic rice plain, the town shares in the slope of the country from Chakai and the Hazaribagh plateau northwards. This circumstance and the proximity of the Kiul ensure excellent drainage and render it one of the healthiest places in the district. To the south of the town is an extensive old fort called Indpegarh, already described in this chapter.

Jamui Subdivision.—The southern subdivision of the district, lying between 24°22′ and 25°7′N. and 85°49′ and 86° 37′E. with an area of 1,303 square miles. At the time of the census of 1901, it comprised an area of 1,593 square miles, and its population was returned at 551,227 in 1901, as compared with 553,917 in 1891. The Sheikhpura thana, which then formed part of it, has since been transferred, with the Barbigha outpost, to the Monghyr subdivision, and the population of the Jamui subdivision, as now constituted is 370,695. This subdivision is the least densely populated part of the district, supporting only 284 persons to the square mile, owing to the fact that in the south it merges in the Chota Nagpur plateau, and contains large areas of jungle. It contains 561 villages, one of which, Jamui, is its headquarters; and four police-stations, viz., Jamui, Chakai, Sikandra, and Jhajha.

Jaynagar.—A small village in the Monghyr subdivision situated close to the Lakhisarai railway station. This place is said to have been the stronghold of a Hindu chief of Magadha, called Indardaun, who, according to local legend, was defeated by a Muhammadan called Makhdum Maulana Nur, whose tomb is at Khagaul, half a mile to the north of the railway station. Indardaun has been identified with Indradyumna, the last of the Pala kings, who was conquered by the Muhammadan invaders under Bakhtiyar Khilji.

The name Jaynagar belongs properly to the position on the south, to which Indradyumna is said to have retired after his defeat by the Muhammadans. The position is formed by two short ridges of small rocky hillocks running parallel from west to east, the opening to the west being closed by an earthen rampart, and that to the east by what were massive works, but are now mere mounds. Between the ridges there are two long parallel mounds, which have every appearance of being—as the people say they are—the ruins of the houses of a street or bazar. On the top of the northern ridge there was once a building of some kind, probably a stupa, and on the southern ridge there are the foundations and part of the walls of a small monastery. The walls have large dressed stones on both faces, and there are quantities of bricks lying about the ruins, as well as on a spur below the monastery, which was levelled to form a terrace for building. To the west and south of the hills there are many fine tanks. According to the people, there are *atharah-ganda-pokhar*, or " eighteen four, i.e., seventy-two tanks ". On the north-west there is a fine sheet of water, upwards of a mile in length from north to south, which has been formed by embankments extending from the western end of the northern ridge. Jaynagar is believed be the most southerly portion of a city, which once extended for four miles along the bank of the Kiul river. General Cunningham points out that it corresponds in position with the Lo-in-ni-lo of Hiuen Tsiang, which lay 200 *li*, or 33 miles, to the west of *Hiranya Parvata*, i.e., Monghyr, and he considers it probable that the Chinese syllables may be only a faulty rendering of Kiul.

Close to the top of the northern ridge the villagers say that Indradyumna had his treasury, which was sealed with a magic seal; but all that can be seen is plain, smooth rock, perhaps artificially smoothed. It is said that Indradyumna had a trusted warrior, whom he raised to the highest posts, until at last he asked for the hand of his master's daughter in marriage. The king was very angry, and had a cavern made in which he placed all his treasure. When all was safely stowed away, the king invited his general to see his treasury, and when he unsuspectingly went in, let fall the trap door and sealed it with a magic seal. It was not long before he suffered for thus killing his best general; for the Muhammadans came down and drove him a fugitive from place to place, until he was obliged to fly to Orissa. His last place of refuge is still pointed out—a natural cavern on the top of the southern ridge. [Reports, Arch. Surv. Ind., Vol. III, pp. 159-160; Vol. VII, pp. 118-119.]

Jhajha.—A railway station on the East Indian Railway situated in the south-east of the Jamui subdivision. The original name of the place is Nawadih, but, to prevent confusion with other places having a similar name, the railway authorities named the station Jhajha. It is a changing station on the line, and the railway company owns a large area of land, on which is located the European and Eurasian staff, numbering with their families over 200 persons. The sanitation, etc., of the settlement is managed by a Station Committee consisting of railway officials. Just outside the railway premises is Nawadih, a crowded *basti* which has one or two fairly broad roads, but otherwise is intersected merely by numerous little gullies and passages. The owner of this *basti* is the Maharaja of Gidhaur. The population is (1921) 3,970 and the place contains a police-station and a district board bungalow situated about a mile away from the railway station.

Kabar Tal.—A lake in the north of the Begusarai subdivision, about eight miles in length and two miles in breadth.

formerly flowed through the north of the district. It is of no great depth in places, and a portion generally dries up in time to allow rice to be sown broad-cast in May, the crop being reaped from boats in November. The water, except where it is very deep, is covered with a thick growth of tall reeds, which give shelter to numerous water fowl, so that the lake is one of the best shooting grounds in the district. It also swarms with fish, and is an important source of the fish supply of the people.

There is an island in the lake known as Monkey Island, from the number of monkeys which frequent it. The island is a sacred spot; and when the Permanent Settlement was concluded in 1793, not only was it allowed to remain revenue-free, but Government made a yearly grant for the express purpose of feeding the monkeys and keeping alight a lamp which burnt day and night in a small temple on the island. This grant was continued until 1852, when it was discovered that the light had been extinguished and the monkeys had not been fed, the money being misappropriated by the priests. The grant was accordingly withdrawn. Palm trees abound on the island, but they are never tapped, for the people believe that they would yield blood instead of toddy.

The temple above mentioned is a small shrine dedicated to Jaimangala, another name for the goddess Durga or Bhawani, a painted figure of whom may be seen in the niche opposite the low door in front. The building is believed to be very ancient, and considerable sanctity attaches to it, pilgrims coming to it from distant parts, especially during the Durga Puja. It is also visited by villagers from the neighbourhood on Tuesdays and Saturdays, these being auspicious days. This is the only temple on the island, which is overgrown with jungle and traversed only by rough tracks. Badly burnt bricks may be found here and there, and a number of cannon balls have been found, so that it is probable that the island was originally a fort. This hypothesis is rendered more probable by the fact that the place is known as Jaimangalgarh, i.e., the fort of Jaimangala. It is situated four miles east of Chiria Bariarpur and 12 miles north of Begusarai.

Khagaria.—A town in the north of the Monghyr subdivision, situated on the river Gandak at a distance of about three miles from its junction with the Ganges. Khagaria is, next to Monghyr and Jamalpur, the largest town in the district, returning a population of 9,521 at the last census. It is an important trade centre, a large proportion of the chillies and grain exported from the north and north-west of the district, and from the Darbhanga district, passing through it. Several Marwari firms have agencies here, and a number of other merchants have settled in the town. There were formerly also many Bengali merchants, but they have been ousted by the competition of the Marwaris and others. The town contains a station on the Cawnpore-Katihar line of the Bengal and North-Western Railway, which passes over a large bridge of seven spans, with a waterway of 650 feet, constructed over the Gandak about 2½ miles to the west. There is also a direct line to Samastipur, which here takes off from the main line. There are here a district board bungalow, thana and dispensary.

Khaira.—A village in the Jamui subdivision, situated five miles south-east of Jamui. It contains the residence of one of the leading zamindars of the district, who belongs to a branch of the family of the Maharaja Bahadur of Gidhaur. The legend accounting for the separation of the two branches has already been given in the article on Gidhaur, and it will be sufficient to state that the founder of the family is said to have been Hari Singh, the eldest son of Puran Mal, Raja of Gidhaur. Hari Singh was kept at Delhi, presumably as a hostage for the loyalty of his father. On the death of the latter, his younger brother was installed as Raja, and, on his return, Hari Singh agreed to accept a portion of the estate. The original seat of the ancestors of both families appears to have been at the foot of the hills near Khaira, where the remains of an old stone fort and of other buildings may still be traced in the jungle. Close by and in better preservation are the remains of a large masonry fort known as Naulakhagarh (q.v.), the erection of which is by local tradition ascribed to Sher Shah, but which may once have been the seat of the family. The late head of the family, Ram Narayan Singh, received the title of Raja Bahadur from Government. Kumar Guru Prasad Singh, son of Raja Bahadur Ram Narayan Singh, sold the estate in January, 1919, to a syndicate of which Rai Bahadur Baijnath Goenka was the principal member. The sale was impeached by the younger brother of Kumar Guru Prasad Singh, on the ground that the estate was inalienable by custom. The litigation has not yet (1925) come to an end; but Kumar Baijnath Singh has failed, in the court of the subordinate judge of Monghyr and in the Patna High Court, in his attempt to have the sale set aside.

Kharagpur.—A village in the Monghyr subdivision situated close to the eastern flank of the Kharagpur Hills, 25 miles south-east of the town of Monghyr and 12 miles from the railway station of Bariarpur. Population (1921) 2,189. Historically Kharagpur is one of the most interesting places in the district. According to tradition, the surrounding country was originally held by an aboriginal race, called Khetauris, who were ruled by 52 chiefs. Three Rajput brothers named Dandu Rai, Basdeo Rai and Mahendar Rai, took service with Raja Sasanka, the Khetauri chief of Kherhi, a small principality ten or twelve miles south-east of Monghyr. In the course of the wars in which they engaged, they collected a large following of Rajputs; and at last, when strong enough, rose in the night on their master, murdered him and his family, and assumed his power. They then entered on a series of successful contests with the neighbouring chiefs, and are credited with conquering all the 52 Khetauri chiefs. The eldest of the three brothers was Dandu Rai, and he passed on the chiefship to his son, Rup Sah.

Coming to historical times, tradition relates that the ruling chief in the time of Jahangir was Sangram Sah, who was summoned to Delhi to do homage to the Emperor. He refused to obey the summons, and his disobedience led to the advance of a punitive force. Sangram Sah toop up a strong position in the defile of Markan; but one of his soldiers went over to the enemy's camp, and agreed to assassinate him in return for a large reward. Coming up while Sangram Sah was praying, the traitor shot him through the head. This act of treachery, however, had little immediate result, for Chandrajot, his widow, so valiantly resisted Baz Bahadur, the leader of the Mughal troops, that, at last despairing of success and wishing to end a contest with a woman, he proposed an armistice. This being agreed to, Baz Bahadur persuaded the Rani that she should go to the Imperial Court and obtain the Emperor's favour, offering to intercede for her. When the Rani arrived at Delhi, her son Toral Mal was thrown into prison, but Baz Bahadur succeeded in obtaining his release.

Toral Mal was induced to change his religion and become a Musalman, being given the name of Rozafzun. The daughter of a noble was given him in marriage, and afterwards a cousin of the Emperor himself. He was advanced to the rank of *mansabdar* of 3,000 horse, while the command of 2,000 horse was given to each of his children, Bihruz Shah and Abdul Shah. Eventually Toral Mal returned to Kharagpur with his wife, leaving Abdul as an hostage at Delhi, and died about A.D. 1635. The Mughal historians give a different account of the career of Toral Mal, or to give him his Muhammadan name Rozafzun, which will be found in Chapter II.

He was succeeded by Bihruz Shah, who according to the chronicles of the family, added to the already large estate, being rewarded for his services in the Kabul war by the grant of *chakla* Midnapore, in which he built a town and named it Kharagpur. Of his successors there is little of interest to record till the time of Muzaffar Ali, who fell under the displeasure of the Nawab, Kasim Ali. A military force being sent to dispossess him, he fled to Ramgarh, whence he was induced to return in the hope of a reconciliation; but on his arrival at Monghyr, he was seized and kept a prisoner. On the downfall of Kasim Ali, Muzaffar Ali obtained his liberty, but did not recover his property, which was placed in charge of an *amil* or manager.

Dr. Buchanan Hamilton gives some further details regarding the history of the family. Muzaffar's son, Faiz Ali, having appealed to the British Agent at Patna, the property was restored to him, but an *amil*, Abu Talib, was left at Kharagpur to protect the interests of the Company. This officer soon accused the Raja of turbulence; and he was again deprived of all authority, his house plundered, and his family thrown into prison, but he himself escaped into the jungle fastnesses of a rebel hill chief, Jagannath Deo.

The Raja now sent an agent to Murshidabad, and complained to Muzaffar Jang, who issued orders to Shitab Rai, the Governor of Bihar, to see justice done. On this, Faiz Ali and the other members of the Raja's family were released from prison, and Abu Talib recalled. That officer, knowing the fate that awaited him, took poison; and his whole wealth was secured by Shitab Rai, who restored nothing to the family, but sent another officer, who allowed the Raja no more authority than before. The Raja then sent his son and his *diwan*, or minister, with another complaint to Murshidabad, but on the way they met Shitab Rai, who sent the son back and persuaded the *diwan* to accompany him to Calcutta. The Raja, believing that the latter had betrayed him, sent another agent, Rudra Mohan by name, to Calcutta, who gave security, and obtained an order that the management of the estate should be restored to the Raja.

About this time the house of the *kanungo* of Kharagpur was robbed, and the *amil* immediately charged the Raja with the crime, which his family of course denied. The charge, however, was believed by Government, and an European subaltern, Mr. Clerk, with two companies of soldiers, was sent to punish him. The Raja retired to the forests, but deputed his son to meet the officer. When the young man came within a day's journey of the troops, some *ghatwals* informed Mr. Clerk that he was not alone, and intended to fight. Thereupon the English soldiers marched by night, and, surprising the party, killed many, but the Raja's son made his escape. After this, Mohan Singh, a Rajput *ghatwal*, informed Mr. Clerk where the Raja was concealed, and the force advancing suddenly caught him and sent him to Patna, where he was put in irons. In 1770 he petitioned against the *ghatwals* and the *amil*. They were called before Shitab Rai, their accusations declared groundless and the Raja released from prison, but ordered to remain at Patna. A few years later (1774—9), Captain Browne appears to have administered the tract in the extreme south as part of the Jungleterry after the Raja had been ousted from his property. In 1781, however, Warren Hastings ordered the estate to be restored to the Raja Kadir Ali.

Kadir Ali was succeeded by Ikbal Ali Khan, after whom came Rahmat Ali, the last of the family who was Raja of Kharagpur. In 1839 he fell into arrears of revenue, in consequence, he alleged, of embezzlements by his agents at Monghyr, who had quite enough money in their possession to pay the Government demand. The whole of his great property, except Haveli Kharagpur, was sold on the 29th January 1840; and Raja Bidyanand Singh of Purnea and Balanath Sahu became the purchasers. In 1845 the remaining estate of the Raja met with a similar fate, being purchased by the Maharaja of Darbhanga. Rahmat Ali Khan died in 1852, and was buried at the entrance of his Imambara. Some descendants of Rahmat Ali are still living in reduced circumstances.

The Kharagpur Raj was at one time a great principality extending from the south of Monghyr to the south of Bhagalpur and the Santal Parganas, and marching with the northern boundary of the territory held by the Birbhum Rajas. Kharagpur itself contains few remains to attest the former greatness of its Rajas. There is, however, a ruined palace built by Raja Bihruz, adjoining which is a three-domed mosque picturesquely situated on the river Man. A marble slab in one of the walls shows that it was built in 1068 A.H., i.e. 1656 A.D. during the reign of the Emperor Shah Jahan. Three miles from Kharagpur there are two temples and two tanks standing opposite to one another on each side of the road leading to Tarapur. The place is said to have been a hunting seat of the Rajas of Kharagpur while they were still Rajputs. The temples and tanks on the south of the road are ascribed to one of the Rajas, and those on the north to Chandravati, one of the Ranis. They are accordingly known as *Rajarani talao.*

Kharagpur is now perhaps best known for a large reservoir constructed fifty years ago, when the Kharagpur estate, with the rest of the property of the Maharaja of Darbhanga, was under the management of the Court of Wards. It is formed by a dam built, two miles west of Kharagpur, across the Man river, which at this point debouches through a narrow gorge in the hills. To the south-west the gorge widens out into a valley hemmed in on all sides by low but abrupt hills, and here a large reservoir has been formed by the accumulation of the river water and of the drainage from the hills and valleys. This reservoir irrigates about 18,000 acres and forms a beautiful lake, which, according to Mr. Lockwood, rivals the renowned lakes of Killarney.

About a mile or two above the dam is a picturesque waterfall, called Panchkumari or the five princesses. In the neighbourhood is a hill also called Panchkumari, a name explained by the following legend. It is said that the five daughters of the Raja of Kharagpur took refuge there when their father was taken prisoner to Delhi. Being unable to escape from their Muhammadan pursuers, they killed themselves by jumping from a precipice, preferring death to dishonour. At the foot of the hill is a spot where the Rani is said to have committed *sati* on hearing that the Raja had been put to death at Delhi.

About 2½ miles south-west of the Panchkumari fall, not far from Karmantari village, is a group of hot springs known as Lakshmikund, which emerge from crevices in rocks on the west side of a narrow torrent bed, some distance above and not far north of the lake, into which they discharge. These springs, which were not noticed by Buchanan, are most conveniently reached from Kharagpur, eight miles to the north-east, by taking a boat across the lake. On the 1st of November, 1917, Mr. V. H. Jackson found that the temperature of eight of the eleven principal outflows was over 146°, while the temperature at the largest was 151·3°, which is higher than the maximum temperature recorded in the Bhimbandh or Janamkund groups, which these springs closely resemble. As at both of the latter groups of springs, there is another series of springs lower down, which are distinctly cooler.

Kharagpur Hills.—A range of hills situated immediately to the south of Monghyr town, and so called after the town of Kharagpur, which lies at the foot of the hills to the east. These hills, which are an offshoot from the northern face of the Vindhya Hills, measure 30 miles in length, with an average breadth of 24 miles. Although the group lithologically resembles the Vindhya Hills, it contains within its valleys, and on some of its higher peaks, rocks of a much softer nature, such as quartzite, chlorite, chlorite schist, hornblende, etc. In the interior are several valleys, precipices, hot springs, hill torrents, quarries and a few villages. The hills nowhere rise to a greater height than 1,628 feet, which is the height of the high table-topped hill 13 miles south of Monghyr, named Maruk. They gradually converge towards Monghyr town, where they dip under the Ganges, in the bed of which there is an out-crop of bare quartz rock. A small hill, named Pirpahar, about three miles east of the station, forms the most northerly point of the hills, which terminate in a perpendicular bluff over-hanging the old bed of the Ganges.*

---

* Captain S. R. Sherwill, *The Kurrukpore Hills*, J. A. S. B., Vol. XXI.

Kiul.—A village in the Monghyr subdivision, situated 34 miles, by rail, from Monghyr, on the eastern bank of the Kiul river. The river is spanned by a fine lattice girder bridge of 9 spans of 150 feet each, and there is a large railway station, Kiul being a junction of the loop and chord lines of the East Indian Railway with the South Bihar Railway. The village is small, uninteresting and modern, owing its creation to the railway ; but the neighbourhood has some interest to the archæologist on account of the remains which have been found. To the south is the small village of Kowaya, which appears to have been built upon part of an ancient city, for the land is a succession of undulating mounds abounding in old bricks and broken images. On the bank of the river, close to the hamlet called Brindaban, there is a conspicuous mound, upwards of 30 feet in height, which evidently marks the remains of a solid brick stupa. It was excavated by General Cunningham, and at a depth of 6 feet, or 25 feet above ground level, a small chamber was uncovered, which contained a relic casket of pale yellow steatite in the shape of a stupa, and a small figure of Buddha of the same material. Inside the casket was a small golden box containing a fragment of bone, and a broken silver box of the same shape and size with a green glass bead. General Cunningham conjectured that the building was of late date, i.e., not earlier than the 9th or 10th century A.D., and this supposition was confirmed by a subsequent discovery. On the eastern face of the mound a small arched chamber was unearthed, and at the bottom were 200 or 300 seals of lac. On the western side a smaller chamber was discovered, in which were four bronze images, a steatite image of Buddha, and a large earthenware jar filled with lac seals, of which altogether 2,700 were counted. These seals on examination were found to be of four different kinds and were presumably the official seals of a monastery located here. The characters of the writing showed that they were engraved in the 10th or 11th century A.D. This mound appears to have been made a quarry for bricks, but a portion still remains, and close by is another mound, which apparently marks the site of an old monastery. General Cunningham has conjectured that Kiul may be an altered form of the name which Hiuen Tsiang has handed down as Lo-in-ni-lo.*

---

* Reports, Arch. Surv. Ind., Vol. III, pp. 156-8.

Lachhuar.—A village in the Jamui subdivision, situated about five miles west of Simaria and four miles south of Sikandra. It contains a large Jain temple and *dharmasala* built in 1874 by Rai Dhanpat Singh Bahadur of Murshidabad, for the benefit of Jain pilgrims, who visit some places in the adjacent hills. The nearest are three miles south of Lachhuar and are marked " Muth Boodhroop " and " Muth Purusnath " on the Indian atlas sheet. They are two small shrines picturesquely situated in a valley between two parallel ranges of hills. In each of these shrines is a small statue of Mahavira, one of which dates back to Sambat 1505, while the other appears to be older. The temples themselves, however, are of recent date.*

Lachhuar is said to have been several centures ago the residence of Puran Mal, Raja of Gidhaur, who built the temples mentioned in the account of Simaria, and of his sons, Hari Singh and Bisambhar Singh. On the outskirts of the village there is a temple dedicated to Kali, which is maintained by the Gidhaur Raj. A large fair is held here during the Kali Puja festival.

Lakhisarai.—A village in the Monghyr subdivision, situated on the western bank of the Kiul river, 35 miles by rail from Monghyr. The village contains the *dargah* or mausoleum of Makhdum Shah, in which is a loose slab of basalt containing an inscription of the Bengal Sultan Rukn-ud-din Kaikaus. The slab is broken into two pieces, and most of the letters of the king's name are lost ; but the fact that the name begins with *Ruknuddunya* and ends with the letter *s*, combined with the date, is sufficient to raise this identification beyond doubt. The only other known inscription of this king, which is inside the *dargah* of Maulana Ata at Gangarampur in the Dinajpur district, bears the same date, viz., the 31st Muharram 697, i.e., 19th October 1297 A.D. The inscription has no connection with the *dargah*, where it now lies, as it refers to the building of a *jama masjid*.

There were also formerly extensive Buddhist remains near Lakhisarai, but very little has escaped destruction. Several brick mounds can be traced along the Kiul river to the south of Lakhisarai, and some years ago numerous ancient

* Report, Arch. Surv. Bengal, 1902-3.

statues are said to have existed there. They have almost all been carried away, and the brick mounds generally have been made level with the ground. Some fine Buddhistic statues found here may be seen in the Indian Museum.*

Malnipahar.—A hill in the Monghyr subdivision, situated in the Kharagpur Hills about seven miles north-east of Bhimbandh. There are several springs, known as Janamkund, at the bottom of the hill, which form the source of the Anjan river. One spring, which issues at all seasons of the year directly from a crevice in the rock, is apparently that of which the temperature was tested by Buchanan in 1811. Mr. V. H. Jackson, making tests at different seasons, has found that its temperature varies from 147·2° to 149°. A second series of springs, the existence of which was suspected by Buchanan, was discovered in 1912. These occur along the bed of the Anjan for about 150 yards, at a quarter of a mile from the source. Their highest temperature yet observed is 140°.†

Maruk.—A hill in the Monghyr subdivision, situated in the Kharagpur Hills 13 miles south of Monghyr. The following account of the hill is quoted from an article on the Kharagpur Hills by Captain Sherwill: "A rough and steep scramble brought us to the summit of Maruk, a table-topped hill of 1,600 feet elevation, from whence we had a splendid view of Monghyr station and town, 13 miles to the north of us; of the country beyond the Ganges for nearly 100 miles; of the Ganges winding through the highly cultivated plains of the districts of Patna, Monghyr and Bhagalpur; a good view of the Rajmahal Hills to the east, distant 70 miles, and of the jungles at our feet. Clouds shut out the view of the Himalaya mountains, which a few days before we had seen from Monghyr in the plains, spread out in a vast panoramic view, their snowy sides tinged with the beams of the rising sun. The summit of this mountain is about a quarter of a mile in length and a few hundred yards in breadth, perfectly level and covered with a matted and tangled jungle of bamboos, mimosa, catechu, and *sakua* trees. The spot from its elevation deserves to have a house or two erected on its summit, where invalids from Monghyr would,

* Reports, Arch. Surv. Bengal, 1902-3.

† Eastern India II, 199; Jackson, Patna College Magazine, Vol. III, 1909, pp. 1—8 and 54—8.

during the great heats of summer, find relief from the difference of temperature. Looking north and down into the jungle, large bare masses of quartz rock are seen protruding through the surface of the country and over-topping the highest trees. The ferruginous clay-like laterite at the summit of the mountain is excavated into natural caves highly polished by the frequent visits of the long-tailed monkeys which abound in the woods in these hills.

"On the summit of this mountain we fell in with several of the gigantic yellow webs of the epeirae spider, which are as remarkable for their strength of web as they are for the variety of their forms and colours. The present specimens are red and black, of a formidable size and very active. Some of the webs we found stretched across our path measuring from 10 to 20 feet in diameter, in the centre of which the spider sits waiting for his prey. The webs from their great strength offered a sensible resistance when forcing our way through them; in the web of one of the spiders we found a bird entangled and the young spiders, about eight in number, feeding upon the carcass. The bird was, with the exception of his legs and beak, entirely enveloped in web, and was much decomposed ; the entwined web had completely pinioned the wings of the bird so as to render his escape impossible. The bird was about the size of a field lark and was near the centre of the web; the old spider was about a foot above the bird ; we secured, measured and bottled him. His dimensions were six inches across the legs ; he was armed with a formidable pair of mandibles."*

The origin of the name Maruk is not known, but it is probably so called after the *maharuk* tree (*Ailanthus excelsa*).

Maulanagar.—A village in the Monghyr subdivision, situated about half a mile east of Surajgarha, six miles north of the Kajra railway station, and 18 miles west of Monghyr. The traditional account of the origin and history of the village is as follows. Formerly the place where Maulanagar now stands was covered by jungle. In the 18th century A.D. a saint, named Shah Nazim-ud-din Ali, happened to come to Surajgarha, where his brother held the post of *Kazi*, and attracted by the solitude of the place, determined to pass the remainder of his life there. He took up his abode below

* *The Kurrukpore Hills*, J. A. S. B., Vol. XXI.

a large tree on the bank of a lake, and people from all quarters began to visit him and beg that he would intercede with God for them. To one and all the saint humbly replied : " Go. Maula (God) will do you good." The people of the neighbourhood, therefore, called him Maula Shah, and the place where he had settled Maulanagar. Here he was visited by Ali Vardi Khan, when he was marching northwards past Surajgarha on one of his frequent expeditions. Hearing of the fame of the saint, Ali Vardi Khan came to him, and, like the poor villagers, begged that he would pray to God for success in his campaign. The saint complied with his request, and, as usual, said : " Go. Maula will do you good." The saint's prayer was answered, and Ali Vardi Khan returning from a successful campaign, made him a grant of two *mahals*, viz., *pargana* Abhaipur and *taluk* Mustaphanagar. This grant he at first refused, but at length consented on condition that its proceeds should be used for charitable purposes.

This tradition is confirmed by the *Sair-ul-Mutakharin*, which mentions Shah Maula among the saints of Bihar, and says that he passed his days in poverty giving his little all to feed passing wayfarers and the poor. " But Haidar Ali Khan, who commanded Ali Vardi Khan's artillery, made application to that Prince for his bestowing on that holy man the small *pargana* of Kajra. The request was granted with the best grace in the world, so that the general one day sent him a *sanad* for it. Now the estate is enjoyed by his descendants, i.e., by the relatives of his wife, who all live comfortably upon the produce of it."* Shah Maula was succeeded by a kinsman and brother-in-law named Shah Ghulam Maula. He was, it is said, a minister of Nadir Shah, king of Persia, and accompanied him in his invasion of India (1739). Instead, however, of following Nadir Shah back to Persia, he came on a visit to Shah Maula, and having become his disciple was appointed his successor. The present Sajjada-nashin of Maulanagar, Saiyad Shah Sami Ahmad, is his descendant.

The family traces back its descent to Saiyad Ahmad, a saint of Medina, and a champion of Islam, and gives the following account of its history. The son of Saiyad Ahmad, Shah Bagh, having been sent by Sikandra Lodi (1488—1517) to punish certain marauders who were ravaging this part of the

* Raymond's translation, II, 175.

country, settled and died in the town of Bihar; and his descendant was Shah Ghulam Maula, the second Sajjada-nashin of Maulanagar. The village contains a *khankah* or Muhammadan religious institution, to which are attached a mosque and *madrasa* and the tombs of the Sajjada-nashins. The property of the endowment consists mainly of *pargana* Abhaipur with an area of 20 square miles, the largest revenue free tenure in the district, which brings in an income of about Rs. 32,000 per annum. Besides this, there are revenue-paying estates consisting of *taluk* Mustaphanagar and Islampur with an annual income of Rs. 8,000. The income is expended in maintaining the *khankah*, mosque and *madrasa*, in feeding the poor and entertaining wayfarers, and in supporting the family and relatives of the Sajjada-nashin.

History.

Monghyr.—Headquarters of the district, situated in 25° 23′N. and 86° 28′E., on the south bank of the Ganges. Tradition ascribes the foundation of the town of Monghyr to Chandra Gupta, after whom it was called Gupta Garh, a name which has been found inscribed on a rock at Kashtaharini Ghat at the north-western corner of the present fort. The town is supposed to have covered an extensive area surrounded with mud walls, remains of which are pointed out as still existing about three miles to the south of the fort, but these seem to belong to a later era. There are, however, unmistakable traces of houses and mosques and other evidence of habitation, extending far beyond the present precincts of the town, which show that even at an early date it must have been a place of much importance. It is mentioned in the *Mahabharata* under the name of Modagiri, and a passage in the Sabha-parva, describing Bhim's conquests in Eastern India, says that after defeating Karna, King of Anga, he fought a battle at Modagiri and killed its chief. The place is known to have been the site of the royal camp of the Pala kings in the 10th century A.D., and inscriptions are still in existence issued from Mudgagiri, as the place was then called, which record the fact that a bridge of boats was built here across the Ganges. During the time of the early Muhammadan kings and emperors, Monghyr town, lying as it does on the banks of the river at a point where the hills close up from the south, and thus commanding both the land and river route, was a position of much importance; and a strong fort, which still exists in excellent preservation, was built to guard the narrow neck of level ground.

At the close of the 15th century Monghyr was the capital of Prince Danyal, son of Ala-ud-din Husain, the Afghan king of Bengal, who in 1497 repaired the fortifications and built a vault over the tomb of Shah Nafah, the Muhammadan patron saint of the town. In 1580 Raja Todar Mal, on being deputed by the Emperor Akbar to reduce the rebellious Afghan chiefs of Bengal, made it his headquarters and constructed entrenchments between the Ganges and the hills. After his defeat by Aurangzeb near Allahabad, Shah Shuja retreated here in 1659, and resolving to make a stand against the imperial troops, strengthened the fortifications and threw up lines of entrenchment; but learning that his retreat was being cut off by Mir Jumla, he hurriedly withdrew his troops from the trenches and beat a retreat to Rajmahal. In the next century the Nawab of Bengal, Mir Kasim Ali, selected Monghyr as his capital and established an arsenal under the supervision of his Armenian General, Gurghin Khan. He retreated here after the defeat of his army at Udhua Nullah, but fled on the approach of the English troops under Major Adams; and the Governor, who was left in command of the fort, capitulated after a two days' bombardment (1763).

The fort was for some time occupied by some of the troops of the East India Company, and in 1766 was the scene of an outbreak among the European officers, known as the White Mutiny, which was effectually quelled by Lord Clive. After this, the fortifications were gradually allowed to fall into disrepair, and it is clear from the account given by the artist William Hodges in 1781 that it was no longer an important position. Hodges visited Monghyr twice in that year—on the second occasion as a guest of Warren Hastings, who was on his way to meet Chait Singh at Benares; but his description of the buildings is, on the whole, so inaccurate, that it is not worth quoting.

Towards the close of the 18th century we find that Monghyr was merely a station for half-pay and invalid officers and men, who formed "a little garrison, sufficient at least for the protection of a depôt of military stores, and of a powder magazine established there. The latter was said to cause sometimes no little uneasiness to the veterans, accustomed as they were to gunpowder. For the most vivid lightning often falls about Monghyr, attracted by the iron ore which abounds in the neighbouring hills, and, if it fell upon the magazine, the whole fort would certainly be destroyed by the explosion."* The place, however, was noted for its salubrious climate, and Warren Hastings speaks in one of his letters of the delightful change of atmosphere from that of Bengal.

Later visitors also wrote in high terms of Monghyr and of its manufactures, but one and all appear to have been much annoyed by the crowds of beggars that infested the place. Bishop Heber, who visited Monghyr in 1823, writes:—"Monghyr, as one approaches it, presents an imposing appearance, having one or two extremely good European houses, each perched on its own little eminence. . . The fort occupies a great deal of ground, but is now dismantled. Its gates, battlements, etc., are all of Asiatic architecture, and precisely similar to those of the Khitairgorod of Moscow. Within is an ample plain of fine turf, dotted with a few trees, and two noble tanks of water, the largest covering, I conceive, a couple of acres. Two high grassy knolls are enclosed within the rampart, occupying two opposite angles of the fort, which is an irregular square, with, I think, 12 semi-circular bastions, and a very wide and deep wet-moat, except on the west side, where it rises immediately from the rocky banks of the river. On one of the eminences of which I speak is a collection of prison-like buildings; on the other a very large and handsome house built originally for the Commander-in-Chief of the district at the time that Monghyr was an important station, and the Marathas were in the neighbourhood; but it was sold some years since by Government. The view from the rampart and the eminences is extremely fine. Monghyr stands on a rocky promontory, with the broad river on both sides, forming two bays, beyond one of which the Rajmahal Hills are visible, and the other is bounded by the nearer range of Kharagpur." Elsewhere the Bishop with his inveterate love of comparing Indian with European places, describes these hills as being "not inferior to the Halkin mountains and the range above Flint and Holywell".†

A quaint account is also given in *Up the Country* by Miss Emily Eden, who came to Monghyr in November 1837. She was met by all the English residents, six in number, "and that is what they call a large station". She was much struck

* T. Twining, *Travels in India a hundred years ago*, 1893.

† R. Heber, *Narrative of a Journey through the Upper Provinces of India*, 1827.

by the inlaid tables and boxes; "and there was the prettiest doll's furniture possible, tables and cane-chairs, and sofas and footstools, of such curious workmanship". Like other visitors, she drove off to Sitakund. "The drive there was a real refreshment. It is the first time for two years I have felt the carriage going up hill at all; and this was not a simple slope, but a good regular hill. Then we came to some genuine rocks—great, bleak, grey stones, with weeds growing between them, and purple hills in the distance. I felt better directly."

Fanny Parkes visited the place in 1836 and 1844, when she wrote :—" The moment we anchored we were assailed with hundreds of beggars; their clamour and cries were most annoying; they were a complete pest; driving them away was useless. The people selling pistols, necklaces, bathing-chairs, baskets, toys, shoes, etc., raised such a hubbub, it was disgusting." She wrote, however, in great admiration of the view from the river :—" The fort is a good object, but on turning to the corner how much was I charmed to see the most picturesque cluster of *bairagi* temples imaginable. The *maths* are surrounded by fine trees; the ruined bastion of the old fort juts out into the river, and has fragments of rock at its base. The high spires of the white temples seen among the trees, the slender bamboos with their bright red or white flags, and a sort of Hindu altar in front, are beautifully grouped. The Directory tells you of the articles in the bazar but omits these gems of oriental beauty, which are invaluable to a lover of the picturesque. Beyond this stretch the walls of the old fort, which are of very great extent, and the view of Monghyr is good from this part of the Ganges. Among the articles manufactured here, the black vases for flowers, turned in white wood, and lacquered whilst on the lathe with sealing wax, are pretty. The necklaces and bracelets in imitation of jet, at two or three rupees the set, are beautifully made; necklaces of St. Agnes' beads, monkeys, chameleons and male bamboos—everything is forthcoming in the bazar, with the exception of ducks. The steamer's passage is from ten to fourteen days to this place—398 miles by the Bhagirathi, 686 by the Sundarbans, and 304 by *dâk*; the latter runs in two days and three-quarters."

A later traveller, Sir Joseph Hooker, describes Monghyr as "by far the prettiest town I had seen on the river, backed by a long range of wooded hills, detached outliers of which rise in the very town. The banks are steep, and they appear more so owing to the fortifications, which are extensive. A number of large, white, two-storied houses, some very imposing, and perched on rounded or conical hills, give a European aspect to the place. Monghyr is celebrated for its iron manufactures, especially of muskets, in which respect it is the Birmingham of Bengal. Generally speaking, these weapons are poor, though stamped with the first English names. A native workman will, however, if time and sufficient reward be given, turn out a first rate fowling-piece. The inhabitants are reported to be sad drunkards, and the abundance of toddy-palms was quite remarkable."

Monghyr did not become an administrative centre till 1812, though the old fort was occupied long before that date by a regiment of the East India Company. At present, it is a purely civil station, and, in some respects, one of the most picturesque in Bengal. It consists of two distinct portions, viz., the fort, within which the public offices and the residences of most of the Europeans lie, and the town outside it stretching away to the east and south. The fort is formed by a great rampart of earth, faced with stone, which encloses a rocky eminence projecting some distance into the Ganges. On the west, the river comes up to the walls and forms a defence on that side; landwards, a deep and wide moat surrounds and protects it. The fort, being built on a foundation of quartzite rock, effectually keeps off any encroachment by the Ganges, but the river oscillates like the pendulum of a clock, taking, it is said, about 80 years between each oscillation. Thus, we find that in 1875 it arrived at its southern limit close to the temple of Chandisthan, and the priests then pointed out the marks made 80 years before by tying ropes on a palm tree imbedded in a *pipal* tree. Old maps show the width of the Ganges as only a mile and a half, but a few years ago it cut away a large expanse of sand near the East Indian Railway station, and now the crossing from the station to the embankment of the Bengal and North-Western Railway branch line to Monghyr Ghat is fully three miles. In September 1908, the Ganges encroached still further, cutting away the river bank near the railway station and forming a deep pot-hole below it. Huts on or near the bank were carried away, and trains had to stop at a level-crossing about a quarter of a mile from the station.

To one entering the fort from the railway station by the main gate, known as the Lal Darwaza or red gate, Monghyr presents a picturesque appearance. The main road runs southwards between two large tanks, behind each of which there is a low hillock. On one of these is the Karnachaura house built by General Goddard, which was originally the residence of the general in command. Subsequently it was acquired by the Maharaja of Vizianagram, and now it is the property of Kumar Kamalaranjan Rai of Cossimbazar. On the other was a fine building, known as the Damdama *Kothi*, which has recently been demolished. Beyond the latter is the palace of Shah Shuja, which has been converted into a jail, and between them lie the Government gardens. The Karnachaura site is a natural rocky eminence in the north-east corner of the fort. The other hillock is an artificial rectangular mound, which was the citadel or acropolis of the fort. Most of the public buildings are also inside the fort, the most important being the civil, revenue and criminal courts, which consist of three parallel rows of buildings. They were built more for utility than for beauty, and mar the effect of the picturesque surroundings. The district board office appears to have been built on the same principle, and the municipal office building is no better. To the west of the courts a trim little ivy-clad church is seen, and to the east of them the English church of the Baptist Mission, built in 1898 to replace a more costly structure erected on the same site in 1863, which was destroyed by the earthquake of 1897. At the eastern gate of the fort is a red brick clock-tower over a horse-shoe arch presented to the town by the late Mr, Herschell Dear.

Near the north gate of the fort is an old cemetery which is full of the obelisk tombs, erected at the end of the 18th and beginning of the 19th century, which Bishop Heber noticed as " almost distinctive of European India ". The earliest of these monuments consists of a massive black stone pillar, probably once part of a temple, which was erected to the memory of a youth named Stewart, who died at Monghyr in 1769, aged 19 years. The next grave is that of John McCabe, Deputy Commissary-General at Monghyr, who died here in 1789. Perhaps, however, the most interesting monument is that of Captain John Williams, presumably the author of *The Rise and Progress of the Bengal Native Infantry*,

He is described as having commanded the invalid battalion of the garrison for many years, and as having died, on board the Hon'ble East India Company's ship *Northumberland* near the Western Islands, on 20th June 1809, aged 68.

The oldest building inside the fort is a Muhammadan shrine built on an elevated piece of ground near the southern gate. It is supposed to contain the grave of a *pir* or saint, whose name is unknown. He is said to have come from Persia to Ajmer and to have directed his steps thence to Monghyr, under instructions from Khwaja Moin-ud-din Chishti, an early Muhammadan missionary in India. The saint lived here for many years and died about the year 596 A.H., corresponding to 1177 A.D. He was buried in an obscure place near the ramparts, and with the lapse of years his burial place was forgotten. Ultimately, in 1497 A.D., when the ramparts of the fort were being repaired by the Governor, Prince Danyal, it was found that every night a portion of the wall fell down, however carefully it had been built during the day. A council of sages was held, and they were all of opinion that there must be the grave of some saint at the place. That night Danyal slept at the place, and had a vision of a *pir* calling upon him to build a mosque over his grave. The prince enquired who he was and how his grave was to be traced. The *pir* refused to disclose his name, but said that his grave could be traced by the smell of musk emanating from it. A search was made, and the grave was easily located. Danyal then caused a mosque to be built over it, and ever since that day it has been known as the *dargah* of Shah Nafah, *nafah* being a Persian word meaning a pod of musk. Over the gateway there is an inscription set up by Prince Danyal, and round the shrine are many old tombs in a dilapidated state.

An interesting tradition also attaches to the Karnachaura house, so called because the highest point of the hill on which it stands is known by the name of Karnachaura, i.e., Karna's seat. These names are accounted for by the following legend of Raja Karna, who is said to be a different person from the well-known hero of the *Mahabharata*, although the story of his liberality proves that he has been confounded with him. Karna of Mudgalpuri (an ancient name for Monghyr) was a contemporary of Vikrama, and an ardent worshipper of the goddess Chandi Devi. Every day he bestowed $1\frac{1}{4}$ maunds of gold on the Brahmans, and every night he visited the shrine of the goddess. There he cast himself into a vessel of boiling *ghi*, and his flesh was devoured by the *joginis*. Pleased with his devotion, the goddess brought the fleshless skeleton to life by sprinkling water over it, and the resuscitated Karna, on rising up, found the *ghi* vessel filled with $1\frac{1}{4}$ maunds of gold. This he bestowed on the Brahmans, and again appearing before the goddess cast himself into the vessel of boiling *ghi*, and was again restored to life by Chandi Devi. At last, the fame of his continued liberality reached Vikrama, who came to Monghyr and became his servant. By close watching, Vikrama discovered the secret of the daily supply of gold, and having one night preceded Karna to the shrine of the goddess, threw himself into the vessel of boiling *ghi*, and being afterwards restored to life, cast himself into the vessel a second time, and yet a third time. His devotion pleased Chandi so much that she told him to ask a boon, and, on his claiming the secret of making gold, she gave him the *paras* or philosopher's stone. When Karna visited the place shortly afterwards, both the goddess and the vessel of *ghi* had disappeared. He then began to sell his property to make his customary gift to the Brahmans, until at last he had nothing left. When Vikrama asked him the cause of his dejection, Karna told him the whole story, and Vikrama at once gave him the *paras* stone. Then Karna thought to himself—" This must be Vikrama, as there is no one else who would be so generous. " So he fell down at his feet in deep obeisance.

Coming to historic times, we know that when the British first occupied Monghyr, they found the remains of a building on the hill and afterwards erected a saluting battery on it. This seems clearly the hill of which the capture by Captain Smith played such an important part in the White Mutiny of 1766, as related in Chapter II. Subsequently, a house was built on it, which William Hodges, accurately enough in an otherwise inaccurate account, mentions as having been built by General Goddard and as being the residence of the Commanding Officer. General Goddard, it may be mentioned, took part in his young days in the White Mutiny at Monghyr. At the close of the 18th century General Briscoe lived in this house, which is referred to as follows by Mr. Twining in *Travels in India a hundred years ago* :—"General Briscoe was the oldest General in the Company's service. He had a noble mansion, situated upon the summit of a small hill within the fort, near the lower angle. It commanded a beautiful view of the river and of the country opposite, and also of the small bay in which our fleet was moored. All boats proceeding up the river were here obliged to cross to the other side, opposite the fort, in order to avoid a bastion which advanced into the river, and opposing the current—here very strong—threw it off with a violence that made it impossible for any boat to pass on this side."

The Damdama *Kothi* already referred to was another old building, which was demolished in order to make room for the residence of the Collector. Hardly any impression could be made on the solid brick walls by ordinary methods, and they had to be blown up by gunpowder, bit by bit. When the debris was removed, numerous holes were discovered, showing the former existence of under-ground rooms. Inside a well in the compound, just above the water-level, two arched passages were found, one leading towards the house, and the other, in the opposite direction, towards the ground now occupied by the jail. The latter occupies one of the finest sites in the town, and consists of a number of detached buildings with a high wall on three sides and the river on the fourth. One of the buildings, now included in the jail and used as a sleeping ward, was the magazine in Muhammadan times, and the building with its massive walls is still intact. The hospital is said to have been the palace zanana, and the godowns include a building believed to have been a small mosque. "In the floor of this mosque," writes Colonel Crawford, "underneath the centre dome, is a dry well or pit, some ten or twelve feet deep. From this well four subterranean passages lead off in different directions. These passages had all been bricked up, a few yards from their entrances, many years before I went to Monghyr. There was a tradition that some prisoners had made their escape from the jail (it is not likely that they ever got out at the other end) along one of these passages, years before. I believe that one of these passages went down to the river bank, which is just outside; a second to a large well in the garden; a third to the subterranean rooms at the Point. Where the fourth may go, I am not prepared to hazard any suggestion; tradition says to Pirpahar, but three miles is rather a tall order for a practicable under-ground passage." *

* Some Notes on Monghyr, *Bengal: Past and Present*, July 1908.

At the north-western corner of the fort is an ancient bathing *ghat* known as the Kashtaharini Ghat. This name means " the bathing place which expels pain, " the tradition being that all people afflicted by grief or bodily pain were at once cured by bathing here. There is a cluster of six temples here which attract crowds of pilgrims during the Rakhi Purnamashi festival. Three of the temples were built about 100 years ago; viz. the shrine of Gangaji, erected by Lalji Sahu, father of Babu Ganga Prasad, a resident of the town, and the shrines of Siva and Rama Janaki both erected by Kanhai Sahu Halwai, also of Monghyr. The temple of Jagannath was built about ten years later by one Baijnath of Bari Bazar, the temple of Radha-Krishna by Babus Bulaki Lal and Ganga Prasad about 50 years ago, and the temple of Lakshmi-Narayan about 40 years ago by Rai Bahadur Kamaleswari Prasad Singh, a public-spirited zamindar of Monghyr. On the wall of the gateway here is an inscription of about the 10th century A.D. which mentions a king Bhagirath and refers to the building of a temple of Siva There is also an old idol with a mutilated nose lying neglected outside one of the temples, apparently of Buddhist type, but having four arms. The view of the *ghat* from the river is charming, particularly on a festival day, when it is crowded with pilgrims in their holiday dresses, or at night, when it is illuminated.

There is another bathing *ghat* which was built fifty years ago by Babu Ram Prasad Das, father of Rai Bahadur Kamaleswari Prasad Singh. This is generally called the Babua Ghat by the masses and Welcome Ghat by the Europeans, and it is used as a landing place by Viceroys and Governors when coming to Monghyr by river. Both names have a quaint origin. The Indian name is due to the fact that Rai Bahadur Kamaleswari Prasad Singh was known popularly as Babuaji, that being a pet name given to the eldest sons of Indian gentlemen. The hybrid name Welcome Ghat is due to the fact that, on both sides of the *ghat*, the word " Welcome " was inscribed in large letters besides an inscription on the face of the archway running "Long life to His Honour".

On the bastion at the western end of the fort along the riverside is the tomb of Mulla Muhammad Saiyad, a poet who wrote under the *nom-de-plume* of Ashraf. The son of Mulla Muhammad Saleh of Mazandaran near the Caspian Sea, he came to India during the reign of Aurangzeb and was employed as tutor to his daughter Zebunnissa Begam, herself a poetess of no mean renown. In 1672 he obtained leave of absence and went to Ispahan, then the capital of Persia, but a few years later came back to India and was employed by Azim-us-Shan, Viceroy of Bihar and the second son of Shah Alam, eldest son of Aurangzeb. In his old age he determined to go on a pilgrimage to Mecca, but did not live to carry out his intention, dying at Monghyr in the year 1704. He was the author of a Masnawi called *Madan Taif;* also of a commentary on the Kafiyat and of a *Diwan.* His complete poetical works are enumerated among the Oudh MSS. The tomb may be seen on the top of the bastion, the lower part of which is used as a kitchen ; and it has no slab or tablet.*

It remains to note the improvements effected in the fort in more modern times. In a letter, dated 1859, addressed to the Collector of Monghyr, the Magistrate, Mr. Birch, speaks of the "unhealthy state of the south and south-west portion of the fort, which is densely crowded with native huts and *kutcha* houses surrounded by low jungle and in the most filthy state. With the exception of a few natives in an official position and a few *amla* attached to the court, the rest of the houses are occupied by a very low class of natives, who are utterly indifferent to the nuisance arising from the accumulation of filth around their dwellings. The conservancy carts cannot remedy the evil, as there are no roads to penetrate the labyrinth of huts ". The Commissioner submitted in 1867 proposals for the improvement of the fort, reporting that Mr. Dear, " a liberal and public-spirited gentleman, resident of Monghyr, who is much attached to the place, and who has at his own expense carried out the measures of improvement ", and an Indian gentleman, Shah Wajid Ali, were " willing to buy out the occupants of the huts over 18 *bighas* 14 *kathas* and 17½ *dhurs* of land, and to build thereon European houses, if a fifty years' lease of the land be granted to them ". This proposal was accepted, and the scheme successfully carried out. Mr. Dear built a number of good bungalows, many of which still exist, and presented to the town the clock-tower over the eastern gate of the fort. The fort is now the most fashionable residential quarter, but the number of houses inside is limited. At present there are 68 residential houses.

* *Bengal: Past and Present,* Vol. II, October 1908, pp. 524-5.

In concluding this account of the fort, reference may be made to its picturesque position and historical associations. These have been well described by Mr. H. Beveridge :—" Few things are more more beautiful or impressive than to sit on a moonlight night on one of the bastions and listen to the Ganges lapping against the foot of the rampart. The great river still flows quietly on, careless whether she be crowded with shipping or is, as she has now become, an almost deserted highway. But how many stirring events have taken place at this promontory since the day when Sita landed at the Kashtaharini Ghat close by, and went to meet her doom at Sitakund. Hindus, Buddhists, Muhammadans and Christians have successively come here, and erected their places of worship, and now all have more or less passed away. It was here that Todar Mal contended with the rebels against Akbar. It was here that Ellis and his companions were brought after the disaster at Manjhi, and before they were taken back to Patna to be massacred. It was from a bastion of the fort, it is said, that the Seth and his faithful servant were flung into the river. It was from the Patna gate that Mir Kasim's wife and huge train of followers set out for Rohtas, when the news came of the defeat at Gheriah."*

The Indian town.

Little is known about the history of the Indian portion of the town. The earliest mention of value in the Collectorate records appears to be in a letter from the Commissioner of the Bhagalpur or 12th Division to the Secretary to the Sadr Board of Revenue at Fort William, dated the 29th May 1850. He writes :—" It appears from the records that the native town and bazar of Monghyr have for a long period (ever since our first occupation of the country) been considered Government property, denominated the Military or Campoo Bazar. This, though constituting one *mahal*, was divided into 12 *tarafs*, viz. (1) Bara Bazar, (2) Deorhee Bazar, (3) Goddard Bazar, (4) Wellesley Bazar, (5) Moghal Bazar, (6) Gorhee Tola, (7) Batemanganj, (8) Topkhana Bazar, (9) Fanok Bazar, *urf* Dalhatta Bazar, (10) Belan Bazar, (11) Rasoolganj and

---

* *Notes of a Holiday Trip to Malda and Bihar*, Calcutta Review, 1891. The writer explains that the legend of the Seths being killed at Monghyr is probably incorrect (cf. Chapter II), but that the unfortunate Ram Narayan, Governor of Patna, was certainly drowned here.

On the fort as it existed in 1670, see *An old description of the Monghyr fort* by Mr. C. E. A. W. Oldham. *Bengal: Past and Present*, Vol. XXVII, part II.

(12) Begumpur Mandaye. " It would appear from the above that the Military Bazar Government estate represents the lands attached to and more immediately adjoining the fort. There have been considerable additions, and the Government estate at the present time extends over a little more than two-thirds of a square mile, or less than an eighth of the entire municipal area (7·6 square miles). There are few buildings in this portion of the town calling for mention, except a large tobacco factory erected by the Peninsular Tobacco Company at Basdeopur, the Diamond Jubilee College, the Zila school, and the hospital, which is built on an eminence facing the fort on the eastern side. There is a small Marwari *dharmasala* close to the Purabsarai railway station, and a more commodious one near the Monghyr railway station was recently built by Babu Baijnath Goenka, a Marwari resident of the town.

The quarter known as Dilawarpur contains the residence of a leading Muhammadan family known as the Shah family. It traces back its descent to Hazrat Maulana Shah Mustapha Sufi, a man of great learning, who was a native of Seistan in Persia. The fame of his learning reached the ears of Akbar, who invited him to his court in Delhi, where he became one of the Emperor's most trusted counsellors. When Akbar marched south to crush the rebellion of the Afghans in Bihar and Bengal, he was accompanied by Shah Mustapha Sufi, who distinguished himself in the field and made it clear that he possessed supernatural powers. Hearing of the holy life led by a saint of Monghyr, called Hazrat Shah Allahdad Arafin, and of the miracles he wrought, he gave up the idea of a worldly career and came to Monghyr to meet the saint. As soon as Shah Mustapha Sufi looked upon the saint, he became insensible, and, when he revived, found himself in possession of divine secrets. He became the disciple of the Hazrat, who made him Sajjada-nashin, and on his death in 1050 A.H. (1650 A.D.) he was buried in Dilawarpur, where his tomb may still be seen. He was succeeded by his son Shah Sharaf-ud-din, to whom Aurangzeb, hearing of his holy life, gave various presents and grants of land His descendants still reside at Dilawarpur.

Population. The population fell from 59,698 in 1872 to 57,077 in 1891, and dropped to 46,913 in 1911. In 1921 it was 46,825. On these figures Mr. P. C. Tallents remarks :—" The population has been falling away since 1872, largely because, apart from the fact that it is the headquarters of the district, there was little to attract outsiders to the town and largely also owing to the destruction of human life caused by plague in recent years. Nowadays there is a strong industrial interest near by in the shape of a tobacco factory on modern lines which employs over 1,500 hands and the force of the plague in the last ten years has been stayed ; perhaps owing to these reasons the population of the town is stationary." The town is, however, an important trade centre, being favourably situated for trade both by rail and river, and it contains a large number of Indian bankers and *mahajans*. Formerly the trade was carried almost exclusively by river, but the greater part has been diverted to the railway. It is connected by a short branch with the loop line of the East Indian Railway and by a steam ferry with the railway system on the north of the Ganges.

Pirpahar.

Three miles east of the town is a hill called Pirpahar, from the top of which a fine view of the surrounding country is obtained. The hill is called after an old Muhammadan saint or *pir*, whose name is no longer remembered, though devotees occasionally come to worship at his grave. There are two old tombs side by side at the foot of the hill, on one of which there is an inscription to the memory of one Mary Anne Beckett, who died in 1832, while the other has a damaged inscription showing, till a few years ago, that it is in memory of a person named D'Oyly : the portion containing the name has now disappeared. The former is somewhat unconventional in form and character, consising of a mausoleum surrounded by four walls open to the sky, and has a memorial tablet inserted in the northern wall, with the uncommon and not unaffecting inscription "Be still, she sleeps". It is not known who Mary Anne Beckett was, but several legends are current about the manner in which she met her death. One is to the effect that she was a young girl who was killed when riding down the hill ; another is that she threw herself down the hill owing to some love trouble ; while another account says that she was the Kashmirian wife of a Colonel Beckett. Nothing is known about the person to whom the other tomb was erected, but Sir Warren Hastings D'Oyly, formerly Collector of Monghyr, to whom a reference was made, states that it is possible that he or she was a relative of a D'Oyly, who was formerly an indigo planter in the district. The inscription which is now obliterated shows that he or she died in 183—, i.e., between 1830 and 1840.

On the top of the hill there is an old house which may be identified with the residence which, according to the *Sair-ul-Mutakharin*, was erected for himself by Ghurghin Khan, the Armenian general of the Nawab Kasim Ali Khan. This is referred to in the *Sair-ul-Mutakharin* as the house on the hill of Sitakund, though the sacred springs of Sitakund are two miles away; and we learn that when Vansittart, the Governor of the East India Company, visited Monghyr in 1762, it was assigned to him for his residence. Thirty years later it appears to have been known as Belvedere and a pleasing description of it is given by Mr. Twining in "*Travels in India a Hundred Years Ago.*" "To-day," he writes, "the Commander-in-Chief and his party dined with General Ellerker at an elegant mansion situated upon the summit of a hill near the river, about two miles from the fort. The name Belvedere, given to this charming villa, was justly deserved by the extraordinary beauty of its situation. The Ganges, escaping from the gorge on which Monghyr stands, assumes the expanse of a lake, bearing on its northern extremity the picturesque battlements of the fort, and bounded to the west by an amphitheatre of verdant hills. The current being thrown on the opposite side by the bastion above mentioned, and by the southern angle of the fortress, all boats ascending the stream keep near the western shore, passing almost under General Ellerker's windows; while the great cotton and other boats, coming suddenly into view from behind the fort, and borne rapidly across the middle of the bay, impart incessant animation to this fine river scene". Former Collectors of Monghyr resided in this house, which commands one of the finest views one can obtain along the Ganges. Both house and hill are now the property of the sons of the late Babu Upendra Nath Mandal of Chandernagore. Close by, on the summit of another small hill, is a house belonging to Babu Ram Lal Mukerji, a public-spirited Bengali gentleman, who placed a large sum at the disposal of Government for the relief of the distressed in times of famine and flood.

Other places of interest.

About a mile from the railway station is a shrine known as Chandisthan, regarding which Buchanan quotes a legend closely resembling that already given above regarding Karnachaura. Another place of interest is a rock in the bed of the river, about half a mile off from the fort, containing a carving on stone representing two feet, which are supposed to be the impression of the feet of Krishna, when he touched the rock in crossing the Ganges. The rock is submerged when the river is in flood, but comes out in winter. It is called Manpatthar, and is one of a group of temple-crowned rocks cropping out from the river bed, which are known as the Beacon Rocks.

Three miles south of Monghyr is a stream called the Dakra Nala, where there may still be seen the massive ruins of a bridge which was blown up during his retreat by Mir Kasim Ali in 1763 in order to retard the pursuit of the British army.* Lines of earth-works, probably thrown up or added to by Shah Shuja, extend from the bank of the Ganges to the hills near Dakra Nala and may still be traced to the south of the town.

Derivation of name.

The derivation of the name Monghyr, or rather of Munger, as it is known in the vernacular, has formed the subject of much speculation. General Cunningham points out that the hill of Monghyr is said to have been called originally Mudgalapuri, Mudgalasrama or Mudgalagiri, after a *rishi* named Mudgala who had taken up his residence on it; and that even as early as the seventh century A.D. the Chinese pilgrim Hiuen Tsiang must have heard of the name, as he relates how a householder, whose stupa was close to the place, was converted by Mudgalaputra, a famous disciple of Buddha. He goes on to say :—" In the Sanskrit inscriptions of the Pala Rajas, the place is called Mudgagiri. As Mudga is the Sanskrit name of the well-known pulse called *mung*, the present name is only a simple contraction of the Sanskrit name. " He adds :— " I have a strong suspicion, however, that the original name may have been connected with the Mons or Mundas, who occupied this part of the country before the advent of the Aryans....It is, however, not impossible that this name may have been derived from the Sanskrit *Muni*, as the hill is said to have been the residence of the Muni Mudgala, and is therefore known as Muniparvata, as well as Mudgalagiri." †

* A. Broome, *History of the Rise and Progress of the Bengal Army* (1850), p. 390.

† Reports, Arch. Surv. Ind., XV, 15, 16, 18.

Colonel Waddell, however, doubts the derivation of the name from *mudga* or *mung* and writes :—" The *mung* bean does not affect hilly or rocky sites, and its cultivation is widespread throughout the Gangetic plain. It seems not unlikely that the vulgar name of the fort, viz. Mungger, is merely a Muhammadan perversion of the old name, somewhat like the change by which Navadwip was converted into Nadia. The fact that the Sanskrit *mudga* can become, in Prakrit, the colloquial *mung*, is almost equally favourable to a derivation from the sage Mudgal, as in ordinary parlance many letters of the old names are elided :—thus, the classical Kashtaharini Ghat close at hand is popularly called Katharini, and the Chinese pilgrim Hiuen Tsiang in the seventh century also mentions the sage Mudgalaputra in connection with this neighbourhood; and the hermitage of his sage here is still a favourable place of Hindu pilgrimage. It seems therefore more probable that the place derives its name from this sage than from the species of pulse called *mung*.* "

An intelligent local Brahman has given the following account of the origin of the name. " In the sixth century after Christ, a Hindu sage, named Madgal Muni, appeared in the city and established two shrines, one at a rock at Kashtaharini, and the other on the rock known as Manpatthar. On the former religious rites were performed during the rainy season, and on the latter, during the dry season. Both these places are still esteemed sacred by the Hindus. In course of time the rock at Kashtaharini was called Madgal Munigir (the rock of Madgal Muni), which was subsequently abbreviated into Munigir and eventually corrupted to Mungir, from which the modern town took its name. " It will be noticed that all these legends agree in saying that a *muni* or sage lived here in olden days; and Mr. C. E. A. W. Oldham, I.C.S., a former Collector, who first brought the above local legend to notice, suggests that it is possible that the original name was Munigriha, i.e., the house or abode of the *muni*—without any specification of his name—that this was corrupted to Mungir just as Rajagriha has been corrupted to Rajgir, and that the modern Munger is a further corruption.

---

* Note on an old inscription from Mungir, Proc. A. S. B., 1890, pp 191-92.

Monghyr Subdivision.—Headquarters subdivision of the district, situated between 24° 57′ and 25° 49′ N., and between 85° 36′ and 86° 51′ E. The subdivision has an area of 1,915 square miles and is divided into two portions by the Ganges. The northern portion is comprised within the police thanas of Gogri, Bakhtiyarpur and Khagaria, Gogri lying to the south-east, Bakhtiyarpur to the north-east and Khagaria to the west. This portion is a low-lying fertile alluvial tract, subject to flood from the rivers which flow through it in a south-easterly direction towards the Ganges. The principal rivers are the Burh Gandak, the Baghmati with its tributary the Chandan, and the Kamla or Tiljuga. Thana Gogri, which is traversed by the Baghmati and Tiljuga, is particularly liable to inundation, and is covered with a network of swamps, which make it one of the best duck-shooting grounds in Bihar. To the south of the Ganges the general level is higher and the surface more undulating, and there are hill ranges, of which the principal is the Kharagpur range. This portion of the subdivision is comprised within the following police thanas :—Kharagpur and Tarapur on the east, Monghyr and Jamalpur on the north, Lakhisarai, Sheikhpura and Barbigha on the west, and Surajgarha in the centre.

The population of the subdivision is 1,096,670, and the density is 572 persons to the square mile It contains 2,108 villages and the towns of Monghyr, Jamalpur, Khagaria and Sheikhpura.

Narhan Estate.—An estate situated in the districts of Darbhanga, Muzaffarpur, Monghyr, and Patna, with an area of 57,282 acres. In Monghyr district 84 villages belong wholly or in part to the estate, forming a compact block in the parganas of Bhusarhi, Naipur and Imadpur. Of these sixty villages, covering an area of 30,359 acres, and containing a population of about 28,500 persons, belong entirely to the estate. The estate is so called because the family residence is at Narhan, a village lying just within the district of Darbhanga, close to the Monghyr boundary. The present proprietress, Rani Rajnit Kuer, was invested with the title of Rani in 1920, in recognition of her activities in carrying out works of public utility, of which the most notable in this district is the Forrest Bridge over the Burh Gandak at Phaphaut.

Naulakagarh.—A ruined fort in the Jamui subdivision situated seven miles south-west of Jamui and three miles south-west of Khaira. The fort, which is picturesquely situated at the foot of the Khaira Hills, is square in shape and is enclosed by thick walls made of unhewn stones and filled with cement. At each corner there is a round tower, and there are four gates, one leading through each wall. The northern entrance appears to have been the principal one, being defended by an out-work. Inside, steps lead up to the top of the walls, and outside, there is a bastion on each side of the four gates. The length of each wall is about 250 feet, but they have fallen down in many places. The fort was probably a stronghold of the Gidhaur Raj when its seat was at Khaira, but popularly it is ascribed to either Akbar or Sher Shah. It is said that after it was finished, the Emperor ordered a cannon to be taken up to the top of a peak in the adjacent range of hills, and, as it was found that the shot fell within the fort, it was abandoned as untenable. The same legend is told about the fort of Shergarh in the south of the Shahabad district, and it is said that the two forts are so similar that they may safely be put down to the same period. There are no traces of buildings inside, and the tradition may therefore be true that the place was given up immediately after it had been built, as it was not considered sufficiently strong. It may have, however, been built merely as a hunting lodge, for it lies just at the foot of several high hills, still covered with thick jungle. The name Naulakhagarh is said to have been given to the fort because nine lakhs of rupees were expended on its construction. Naulakha is however a common name for anything big, e.g., a mango grove supposed to contain nine lakhs of trees, the Naulakha buildings at Bhojpur supposed to contain nine lakhs of bricks, etc. (Reports, Arch. Surv. Ind., Vol. VIII, 1878; Report, Arch. Surv. Bengal Circle, for 1902-3.)

Nongarh.—A village in the Jamui subdivision, situated on the west bank of the Kiul river about eleven miles south-east of Lakhisarai. The village derives its name from a great mound called Nongarh, which is 40 feet in height and 200 feet in diameter at the base. It is a solid mass of well-burnt bricks, each 12 inches by 9 inches by 2 inches and was evidently an ancient Buddhist stupa. It was excavated by General Cunningham, who sank a shaft from the top downwards and discovered, seven feet from the top, a small chamber containing three small clay stupas, and below that another chamber with eight more stupas of clay, besides a broken statue bearing the remains of an inscription in early characters of the first century before or after Christ. The statue is made of the red-spotted sandstone of the Sikri quarries near Mathura, and the treatment of the drapery, fitting close to the figure, is the same as that of the Mathura statues of the same age. Excavation was continued by Mr. Beglar, who unearthed, 19 feet lower down, some fragments of an arch of brick, built edge to edge, and an even brick floor, which appears to have been the floor of the sanctum of a small temple.

The results of his exploration are described by Mr. Beglar as follows :—" It would appear that there once existed here a small temple facing north; that in course of time this temple fell to ruin and became a low mound above 12 or 13 feet high; and that subsequently on this mound a stupa was built. There is nothing to show the age of the stupa beyond a small model stupa found by General Cunningham in the relic chamber. Judging from this, it is not probable that the stupa is so old as the first century before or after Christ; but the existence of the mutilated red-stone statue, with its inscribed characters, shows that there was some sort of religious building here as early as the beginning of the Christian era. As it is clear that the stupa was built on the ruins of the temple below, and as the stupa is clearly not of the period about the beginning of the Christian era, the temple on whose ruins it stands must be considered of the same age as the inscribed red-stone statue, viz. of the first century before or after Christ. Having no doubt myself that the temple was certainly as old as the statue, which, according to General Cunningham, dates to the first century before or after Christ, it follows—(1) That the true arch was known and used in India at that time. (2) That although the principle of the true arch was known, it was, so far as yet positively known, built invariably of bricks edge to edge, and not face to face as our modern arches. (3) That the use of mortar, lime and *surki* was known. (4) That fine lime plastering was known and used at that early period. From the mutilated statue it appears to me idle to speculate as to the deity to whom the temple was dedicated, whether Buddhist or Brahmanical; the probabilities are in favour of its being Buddhist."

About 200 feet to the east of the stupa there are the remains of a monastery, of which about half has been carried away by the river. General Cunningham was at one time inclined to identify Nongarh, also called Longarh, with the Lo-in-ni-lo or Lonyara of Hiuen Tsiang, as the names are very nearly the same, while the only two buildings which the pilgrim mentions, a monastery and a great stupa, correspond with the only two ruins now existing at Nongarh. In these respects Nongarh corresponds with the Lo-in-ni-lo of Hiuen Tsiang; but he also mentions a large lake, and as this no longer exists, its correspondence with the pilgrim's description is imperfect. Elsewhere General Cunningham identifies Lo-in-ni-lo with Kiul. [Reports, Arch. Surv. Ind, Vol. III, pp. 160-2, and Vol. VIII, pp. 118-20.]

Pharkiya Pargana.—A pargana in the north-east of the Monghyr subdivision, with an area of 506 square miles, comprised mainly within the Gogri thana. This tract formerly belonged to an ancient family of zamindars, of whose history little is known except what was collected in 1787 by Mr. Adair, the Collector of Bhagalpur. It is said to have been inhabited by a lawless tribe of Hindus of the Dosadh caste, who made constant depredations in the neighbouring country. At last towards the close of the 15th century the Emperor of Delhi sent a Rajput, named Biswanath Rai, to restore order, a task which he successfully accomplished. He then obtained the grant of a zamindari in this part of the country, and the estate devolved on his posterity without interruption for ten generations. The annals of the family, however, after the first quarter of the 18th century, are a record of little but bloodshed and violence, affording striking evidence of the state of Government and society that preceded the English conquest of Bengal. In 1730 Kunjal Singh, who then held the estate, was treacherously murdered by a Chakwar, named Ruko Singh, who plundered the country. Next year Narayan Dat, the brother of Kunjal Singh, obtained possession, but was soon after confined for arrears of revenue in the defaulters' prison at Patna; and in his absence Paspat Rai and Rupnarayan,

Rajas of Tirhut, appear successively to have had possession of the *pargana*, but whether by authority or force is uncertain. Narayan Dat at length obtained his release and re-established himself in the zamindari, which he held until A.D. 1742 when he was killed at Patna by one Izzat Khan, who seized his property.

A short time afterwards, Bhawani Singh of Kharagpur defeated Izzat Khan in battle, and took possession of the property. He set up a right of inheritance through descent from the original grantee, and leaving his brother Kalian Singh in charge, resided at Patna. Hardat Singh, the legitimate descendant of Narayan Dat, denied this pretension, and in 1757 submitted his claims to Waris Ali Khan, *amil* of Bhagalpur. Before the case could be brought to issue, news arrived of the recall of Waris Ali; and the same night Hardat Singh, fearing that the appointment of a new *amil* might prove an obstacle to his suit, assembled his adherents, and having procured admission into the defaulters' prison, where Kalian Singh and his brothers were confined, put them all to death, and took possession of the property in dispute. Although such a proceeding was liable to punishment, even under the Mughal Government, Hardat Singh had no hesitation in avowing his crime; and without being called to any account for it, was suffered to remain in possession of the zamindari until A.D. 1766. when Waris Ali Khan, being again *amil* of Bhagalpur, drove him out to make room for Mukam Singh, a relative of the murdered Kalian. Next year, however, it was thought proper to dispossess the latter and restore Hardat Singh, who continued in possession down to A.D. 1790.

In the early years of the 19th century one Buniad Singh was the principal zamindar of this *pargana*, having, it is said, 176 villages covering an area of 74,038 *bighas* and assessed to a revenue of Rs. 17,432. The whole of this estate was sold up for arrears of revenue in 1798, but the price obtained did not cover the amount of the arrears. Buniad Singh was accordingly put in jail, and, so far as can be ascertained, he remained there for several years. In the meanwhile, Bhuttan Singh, who had purchased a portion of the estate assessed at Rs. 12,126, was also sold up in 1802, and another speculative purchaser took his place. The only property left to Buniad Singh consisted of waste lands not assessed to revenue but included in this settlement; their lands had been put up to sale, but no one would bid for them. The reason for this failure to pay land revenue appears to be that it was a matter of no little difficulty for the zamindar to collect his dues in such an area of swamp and jungle, while his position was not improved (by the fact that the estate included large areas on which the only rent realizable was a tax on animals, that were brought then, as now, to graze in the jungle and prairie grasses. Further details of this *pargana* will be found in Chapter X. The greater portion of the estate is now the property of Babu Kedarnath Goenka and Babu Deonandan Prasad.

Pirpahar.—*See* Monghyr.

Rajaona.—A village in the Monghyr subdivision situated two miles north-west of Lakhisarai. This village has been identified by General Cunningham with Lo-in-ni-lo, a place visited by the Chinese pilgrim Hiuen Tsiang in the 7th century A.D., which possessed a monastery and stupa erected by Asoka, with a lake 30 *li*, i.e., five miles in circuit, lying 2 or 3 *li*, i.e., rather less than half-a-mile, to the north of the stupa. "The only place which suits this special description of Lo-in-ni-lo is Rajaona, which is situated near the junction of the Kiul river with the old Ganges or Halahar. It still possesses a large sheet of water to the north, which is supplied by the overflow of the Halahar, direct from the Ganges. The position must have been a favourite one, as the mounds of ruins showing the foundations of both Buddhist stupas and Brahmanical temples extend for four miles along the western or left bank of the Kiul river with a varying breadth of from 1 mile to 1½ mile. At the northern end is the large village of Rajaona, and at the south the fortified hill city of Jaynagar, with a lake to the north-west, about three miles in circuit. Between Rajaona and the railway station is the small village of Khagol, which possesses the *dargah* of a great saint named Pir Makhdum Maulana Nur, who is said to have defeated Indardaun or Indradyumna, the last Raja of Jaynagar."

Rajaona is a village surrounded by numerous mounds and was one of the *mahals* of *Sarkar* Monghyr in the time of Akbar. The ruins have furnished several miles of brick ballast to the railway. There used to be numerous

Buddhist statues and sculptures here, but nearly all have been carried off to different temples and the Indian Museum. Some statues still exist at an ancient mound called Raghugarh, viz., Ganesa, Harihara, Durga, Vishnu, the seven mothers, the nine planets, etc., as well as a few imperfect Buddhist figures, but most of them have been taken away. [Reports, Arch. Surv. Ind., Vol. III, pp. 151—6 and Vol. XV, pp. 13—15.]

Rameshwara Kund.—A hot spring rising at the foot of a hill near the Panchkumari waterfall (see Kharagpur) and flowing into the Man river. The temperature of the spring varies from 110·2° to 111·2°. The spring appears to resemble those of Rishikund in all essential respects.

Rishikund.—A hot spring in the Monghyr subdivision situated about six miles south of Sitakund at the head of a picturesque little valley between two ridges of the Kharagpur Hills. It has been made a place of worship, and a reservoir, about 140 feet square, has been built to collect the water. The bottom is in some places sandy, in others rocky; and the water seems to issue all along the western side from numerous crevices in the rock. Bubbles rise from the whole extent of the pool near the hill, and where the gas issues from among sand, it forms cavities like minute craters. According to observations taken by Buchanan on the morning of the 8th April, 1811, the thermometer in the air stood at 72°; in the water, where it issued from the crevice of a rock, it rose to 110°, and in one of the cavities to 114°.

Sheikhpura.—A village in the extreme south-west of the Monghyr subdivision with a station situated on the South Bihar Railway. Population (1921) 11,937. It is an important centre for the grain trade and for the manufacture of hookah tubes, and contains a district board bungalow, police-station, and dispensary. Sheikhpura has been identified by General Cunningham with a village visited by the Chinese pilgrim Hiuen Tsiang in the seventh century A.D. Hiuen Tsiang, after leaving the Gaya district, arrived at a large and populous village to the south of the Ganges, which possessed many Brahmanical temples ornamented with fine sculptures. There was also a great stupa built on the spot where Buddha had preached for one night. "Both distance and direction point to the vicinity of Sheikhpura, a position which is confirmed by the subsequent easterly route of the pilgrim through forests and gorges of mountains." There are very few ancient remains except a fine tank, two miles west of the village, called Mathokar Tal, on the bank of which there is a *dargah*, said to be the tomb of one Mathokar Khan. But as the site is said to have been originally occupied by a temple of Kali, and as the tank is still called Kali Mathokar, the name is probably only a contraction of Mathpokhar, or the temple-tank, the full name having been Kali-math-pokhar, i.e., the tank of the temple of Kali.*

About three miles to the east, near a place called Pachna, there is a pass over the hills called Goalinkhand to which an interesting legend attaches. The Emperor Sher Shah, it is said, was always fond of Monghyr because it was there that he obtained an early success which formed a stepping stone in his career. Once when marching to quell a rebellion in Bengal, he stopped a week in the fort during the month of Baisakh, the best time of the year for hunting. The Governor had made preparations for a hunt in the jungles near the Sheikhpura Hills; and much to the surprise of his courtiers, the Emperor, on coming to the line of elephants drawn up, ordered the *mahaut* to give him the reins. The astonished *mahaut* replied that an elephant was guided not by reins but by an *ankus*, whereupon Sher Shah, jumping down, mounted his horse and rode off. The courtiers were astonished, and while some admired his courage in wishing to control an elephant by reins, others exclaimed at his whimsical temper, while others murmured that the *jagir* of Sasaram could still be smelt through the perfume of the throne of Delhi. In the meantime, the Emperor was wandering by himself in disguise, making the acquaintance of his subjects like Harun-ul-rashid. Among others he met an old *goalin* or milkwoman of Sheikhpura, who watered his horse, and gave him milk to drink and some pulse to eat. While he was conversing with her, one of his followers, Mian Sulaiman, who had been searching for him, came up and addressed him as Emperor. He asked her what he could do for her to repay her kindness, and she then replied that the best thing he could do would be to make a straight road over the hills to save her and the villagers from the tedious track round them. The Emperor

*Reports, Arch. Surv. Ind., Vol. XV, pp. 12-13.

promised to make a road, and was as good as his word. He would not, however, let it be named after himself, but called it Goalinkhand, or the milkwoman's road.

Sikandra.—A village in the Jamui subdivision, situated 13 miles west of Jamui, 18 miles south-east of Sheikhpura and 15 miles south-west of Lakhisarai. It contains a police-station and district board bungalow, and also the *dargah* of a saint called Shah Muzaffar. This is an ordinary brick tomb in a small open enclosure inside the courtyard of a mosque. The legend is that Shah Muzaffar was king of Balkh in Turkistan, but gave up his kingdom and became a *fakir*. He came to India and made his way to Bihar, where he became the disciple of a famous saint called Shah Makhdum Sharif-ud-din. The latter ordered him to go to Sikandra, where there was a terrible demon, who every day devoured one of the villagers. This demon he subdued, and then performed a *chila*, i.e., remained fasting in fervent devotion for 40 days. Sikandra was formerly the head-quarters of the Jamui subdivision, and a number of roads converge upon it.

Simultala.—A village and railway station on the Chord Line of the East Indian Railway, 217 miles from Calcutta. It is just over 1,000 feet above sea-level, and its healthy climate and picturesque surroundings have made it a popular health resort with Bengalis. It is situated on undulating ground, partly gravel and partly sand, and owing to the slope is rapidly drained. To the north and west are pretty little hills of diverse shapes; and the climate is always cooler than in the low alluvial lands of the district. The following history of the place is quoted from a report sent by the District Officer. Until 1894 Simultala was an obscure village and, besides the railway quarters, contained only one bungalow, the property of Babu Sarat Chandra Mitra, grandson of the late Babu Ram Chandra Mitra, Professor in the Presidency College. This bungalow appears to have been built by a Mr. Smith, who had a mica business in the neighbourhood, and after the death of Mr. and Mrs. Smith came into the possession of the zamindar, Thakur Ranjit Narayan Singh of Telua, from whom Babu Sarat Chandra Mitra acquired it over 20 years ago. In 1894 Mr. Surendra Nath Banerji, editor of the *Bengalee*, came with his family to Simultala and occupied this bungalow. He saw the advantages of the place as a sanatorium and health resort; and next year two Bengali gentlemen persuaded the Thakur of Telua to grant a *mukarari* lease of the elevated site near the railway station known as the Ridge; others soon followed suit and acquired sites. The first masonry house was built in 1897, and a number of others were erected in the next ten years. At present there are about 50 substantial houses in the station, all belonging to Bengali gentlemen, and a few are being added every year. A charitable dispensary was opened twenty years ago, chiefly through the exertions of Mr. S. P. Sinha (now Lord Sinha) and is maintained by public subscriptions. There is a post office, and a mosque for Muhammadans also built by public subscription. Simaltala has a wide reputation among the Bengali community as a sanatorium for the cure of malarious fevers and diseases of the lungs, and many people come here for a change, the season beginning about the month of October and continuing till the close of the cold weather. There is no bazar at the place, and residents and visitors have to get their supplies from the *hats* held twice a week at Telua three miles off, or from the bazar at Jhajha or from Calcutta.

Simaria.—A village in the Jamui subdivision, situated seven miles west of Jamui and about half-way between that place and Sikandra. Close to the Jamui-Sikandra road is a group of six temples in a rectangular compound surrounded on three sides by the water of a large tank. The principal temple enshrines a linga, while inside the minor ones there are, among other images, several Buddhistic statues. One, a large statue of Buddha, has an inscription on its pedestal containing the name of the donor, one Thakkura Buddhasena, and the Buddhistic creed very ungrammatically written. The temples are maintained by the Maharaja Bahadur of Gidhaur, and are said to have been built some centuries ago by his ancestors. There is an interesting legend connected with the largest and oldest, which was built by Raja Puran Mal. Puran Mal, who lived at Lachhuar five miles to the west, was a faithful servant of the god Mahadeo and every day used to ride to Baidyanath to worship in his temple. The god, pleased with his devotion and wishing to save him the trouble of going every day to Baidyanath, appeared to him in a vision and told him that he would find an emblem of his divinity, in other words, a linga, under a potter's wheel in the village. Over this he was to build a temple, at which worship would be as effective as at Baidyanath. The Raja unearthed the linga, and the temple which he built over it was called Dhaneswarnath or Mahadeo Simaria. This legend explains the fact that the officiating priests at the temple are not Brahmans but members of the low caste of Kumhars or potters. According to their own account, the linga Dhaneswar was unearthed by their ancestor. The Buddhist images are now worshipped as Lakshmi Astabhuja, Parvati, Bhairo and Sandhya. Mahadeo Simaria is considered to be a place of considerable sanctity, and pilgrims on their way to Baidyanath make a halt there to bathe the linga with water from the Ganges, which they bring for that purpose.

Sitakund.—A village in the Monghyr subdivision, situated four miles east of the town of Monghyr. It contains a hot spring known as the Sitakund spring, which is so called after the well-known episode of the *Ramayana*. Rama, after rescuing his wife Sita from the demon king Ravana, suspected that she could not have maintained her honour intact, and Sita, to prove her chastity, agreed to enter a blazing fire. She came out of the fiery ordeal unscathed, and imparted to the pool in which she bathed the heat she had absorbed from the fire. The hot spring is now enclosed in a masonry reservoir and is visited by large numbers of pilgrims, especially at the full moon of Magh. The water is beautifully clear and limpid, and sends up numerous bubbles from its rocky bed. The temperature of this spring varies in a remarkable manner, just as described in 1765 by Tieffenthaler :—" The water retains its great heat for about eight months; from the vernal equinox to about the summer solstice, which is the season of greatest heat in these countries, it is less burning and becomes lukewarm."

The earliest exact measurements were made in 1811 by Buchanan who wrote :—" I visited this spring first on the 7th April, a little after sunrise. The thermometer in the open air stood at 68°F., and in the hottest part of the reservoir where many air bubbles rose, it stood at 130°. The priests said that about eight days before it had become cooler, and that the heat would gradually diminish till the commencement of the rainy season. I visited the spring again on the 20th April at sunset, the air having been hot all day and parching; the thermometer in the air stood at 84°, in the well it rose to 122°. On the 28th April I visited it again a little after sunset, the wind blowing strong from the east, but not parching. The temperature in the air was at 90°; in the well it only rose to 92°. The water still continued clear; but soon after, owing to the reduction of the heat, and the natives being in consequence able to bathe in the well, the water became so dirty as to be no longer drinkable by an European. Indisposition for some time prevented me from being able to revisit the place; but in the beginning of July, on the commencement of the rainy season, the water, in consequence of the return of the heat, became again limpid; and on the 26th of that month a native sent with the thermometer found at sunset that it stood in the air at 90°, and in the water at 132°. In the evening of the 21st September, the thermometer stood in the air at 88°, in the cistern at 138°, and the number of air bubbles had very evidently increased." When Sir Joseph Hooker visited the place on April 1st, 1848, he found the temperature to be only 104°, and Colonel Waddell recorded a drop from 137° to 136° between the 11th of January and the 5th of March in 1890. He adds :—" I find on enquiry from the priests at Sitakund that the water still becomes slightly cooler in early summer, but since forty years ago it has never become so cool as to permit of bathing, and they endeavour to make a miracle of this by saying that the annual cooling of the pool ceased immediately after the visit of a certain Mahratta Raja."*

Various explanations of this phenomenon have been suggested, such as "deep-seated thermo-dynamic action", and variations of underground volcanic activity, or of the conductivity of the earth's crust when dry or when soaked with rain; but Mr. V. H. Jackson considers that the real reason is simply the variation of flow of the spring itself during the rainy season and for some months afterwards, the spring flows sufficiently strongly to keep the whole volume of water in the reservoir nearly at its own maximum temperature, which is about 139°. Later on, as the influence of the monsoon wanes, the spring noticeably slackens off, and the water in the tank naturally cools more and more, until the flow is renewed by the next rainy season. During the hot weather following an unfavourable monsoon, the spring may practically dry up altogether. This probably happened in 1811, judging from Buchanan's observations, and it certainly occurred after the failure of the rains in 1908, during the hot weather of 1909, when the reservoir was cleaned out and repaired.

* Some Hot Springs in South Bihar, J. A. S. B., 1890.

Close to the Sitakund spring there is a Hindu temple; and to the north is a reservoir of cold water known as the Ramkund; while to the west there are three more pools called, after the three brothers of Rama, Lakshmankund, Bharatkund, and Satrughnakund. About 300 yards north-west from Sitakund there is a spring on the bank of a pond in the Muhammadan village of Barde; but it is not worshipped, and it is only visible as a surface spring in autumn and winter. Observations taken in January showed that it had exactly the same temperature as Sitakund, viz., 137°F.; but in March no spring was visible, and on digging down two feet the temperature of the water only registered 103°F. There is another hot spring about one-third of a mile south-east from Sitakund, which emerges at the base of the small quartzite hill of Bhainsa, at its southern end. When visited by Colonel Waddell in March of 1890 it was a sluggish spring, with a temperature of 102°, but in October of 1917, Mr. V. H. Jackson found three small pools, the hottest being at 118·7°. This spring is usually submerged under two or three feet of water during the rainy season. The water is only drunk by cattle and no sulphuretted smell is perceptible. Yet another hot spring was found close to Sitakund about thirty years ago and is named Phillipskund after the then Collector, Mr. Phillips. From this spring Messrs. Kellner and Company obtain their supply for the manufacture of ærated water. An observation taken in February 1908 showed the temperature of the water to be 131° or only 1° less than that of Sitakund. In September of 1909 and October of 1912 and 1917 its average temperature was 134·7°, or 3·5° lower than that of Sitakund.

Early European travellers have left interesting accounts of the Sitakund spring. Mr. Twining, who visited the place in 1794, remarked that the water was often sent down to Calcutta for the use of persons about to undertake a long sea voyage, and that if put in bottles it would preserve its tasteless purity for more than twelve months. Bishop Heber also stated that some persons in Calcutta drank nothing else, while Sir Joseph Hooker observed that "the water, which is clear and tasteless, is so pure as to be exported copiously, and the Monghyr manufactory of soda-water presents the anomaly of owing its purity to Sita's ablutions". An interesting account is also given in the *Wanderings of a Pilgrim* by Fanny Parkes, who came here in 1836. She describes how, a few years before her visit, an artilleryman attempted for a wager to swim across the basin, and although he succeeded in getting over, it was necessary to convey him to a hospital, where he died within a few hours from the effect of the hot water.

Sringirikh.—One of the peaks of the Kharagpur group of hills, situated 20 miles to the south-west of Monghyr. The hill is named after the famous Rishya Sringa of the *Ramayana*, who performed a *yajna* sacrifice at the instance of King Dasaratha in order that the latter might have offspring. It is a much frequented place of pilgrimage, especially on the Sivaratri day in February. There is a spring here in a gorge among the hills, which issues in six or seven places from below a high cliff of quartzite and forms a considerable stream lower down. A small reservoir has been constructed at the foot of the cliff, and is used for bathing. It is believed to have miraculous properties, the story being that whoever goes into it, whether child or adult, short or tall, finds the water only waist-deep. The water is hardly lukewarm. Colonel Waddell recorded 90·5°F. in January of 1890, whereas Mr. V. H. Jackson found the temperature to be 86·7° in March and 87·1° in October of 1909. There is also a temple dedicated to Mahadeo, a small square structure, about 15 feet high, with a pyramid over it. It is said to have been built about thirty years ago, by a Marwari, to whom children were born after he had worshipped here. The emblem of the deity enshrined in the temple is an ordinary linga brought from Benares by this Marwari. Another linga lying outside is said to have been the image originally worshipped. Several years ago, so the story goes, a madman removed it from the temple and threw it into a stream, and it was discovered only after a long search. Near it is a female figure, about four feet high, carved in relief on black stone, holding in her hands two long flowers which give support to two small elephants. Two smaller female figures, carved on the same block, stand at the two lower corners on either side of the bigger image. These images are probably Buddhistic, but are now worshipped by Hindus, the bigger image as Parvati, the smaller ones as Gaura and Sandhya. General Cunningham states that he found several figures here, both Buddhistic and Brahminical and two inscriptions, one of which was Buddhistic.* The temple is about six miles from Kajra railway station, but is more easily accessible from Mananpur railway station, eight miles to the south-west.

* Reports, Arch. Surv. Ind., Vol. XV, pp. 19-20.

Surajgarha.—A village in the Monghyr subdivision, situated on the southern bank of the Ganges six miles from the Kajra railway station, and about 25 miles from Monghyr. The place is believed to be one of the oldest in the district, and tradition stated that it is so called because it contained the fort of Raja Suraj Mal, who ruled until the Muhammadan conquest. A decisive battle was fought close by in 1557 between Bahadur Shah, King of Bengal, and Muhammad Shah Adli; Professor Blochmann has located the battle-field at the village of Fatehpur four miles to the west. Later, we find that Ali Vardi Khan halted here, and it is said that the place was the headquarters of a Muhammadan Kazi. There are no remains, however, of any interest, as the Ganges has swept away a large portion of the old village. There was formerly an old fort, but only a small portion of its enclosure is now left. It is reported that about 60 years ago the encroachment of the river laid bare an underground room with a shelf, on which were some old turbans, that crumbled into dust when touched. Surajgarha also used to be an important emporium, but most of its trade has been diverted by the railway. At the foot of a tree close to the ferry *ghat* are several images, both Brahminical and Buddhistic. One is a big Siva linga evidently of recent date ; another is a figure of Buddha sitting. Two are images carved in relief on black stone of some god resembling in some respects Buddha, but holding in his four hands the *sanka* (conch), *chakra* (disk), *gada* (club) and *padma* (lotus), which are usually found with the Hindu god Narayan. One is about three feet high, the other is about half that height. The bigger of the two is surmounted by what is known as a *chalchitra,* in which are carved figures of some animals and birds; and near the bottom of the smaller one are two female figures, one holding a chowrie, the other a guitar.

Uren.—A village in the Monghyr subdivision, situated close to the railway three miles west of Kajra station. It contains several Buddhistic remains, which were first discovered by Colonel Waddell, who identified the site with the place where Buddha converted a Yaksha king called Vakula. It is described as follows by Hiuen Tsiang :—" On the western frontier of the country of I-lan-na-po-fa-to, to the south of the river Ganges, we come to a small solitary mountain with a double peak rising high.* Formerly Buddha in this place

* Beal notes that the passage might be translated "There is a small solitary hill with successive crags heaped up."

rested during the three months of rain, and subdued the Yaksha Vakula (Yo-c'ha Po-khu-lo). Below a corner of the south-east of the mountain is a great stone. On this are marks caused by Buddha sitting thereon. The marks are about an inch deep, five feet two inches long, and two feet one inch wide. Above them is built a stupa. Again to the south is the impression on a stone where Buddha set down his *kiun-chi-kia* (*kundika* or water vessel). In depth the lines are about an inch, and are like a flower with eight buds (or petals). Not far to the south-east of this spot are the foot traces of the Yaksha Vakula. They are about one foot five or six inches long, seven or eight inches wide, and in depth less than two inches. Behind these traces of the Yaksha is a stone figure of Buddha in sitting posture, about six or seven feet high. Next, to the west, not far off, is a place where Buddha walked for exercise. Above this mountain top is the old residence of the Yaksha. Next, to the north is a foot-trace of Buddha, a foot and eight inches long and perhaps six inches wide and half an inch deep. Above it a stupa is erected. Formerly when Buddha subdued the Yaksha, he commanded him not to kill men nor eat their flesh. Having respectfully received the law of Buddha, he was born in heaven. To the west of this are six or seven hot springs. The water is exceedingly hot."

Colonel Waddell claims that Uren satisfies the above description. After showing that its position corresponds with that given by the Chinese pilgrim, he writes—"Of the hill itself no more concise description could be given than that contained in Beal's translation, viz., a small solitary hill with successive crags heaped up." The hill is also "a small solitary mountain with a double peak rising high". "In appearance, the hill literally satisfies both the original and alternative descriptions." As regards the details mentioned by Hiuen Tsiang, Colonel Waddell points out that there is a tradition that the hill was formerly the abode of Lorik, famous in the folklore of Bihar, and identifies the residence of the Yaksha Vakula with what the villagers call his house (*Lorik ka ghar*). This is a somewhat flat area on the top of the hill, below the south-east side of the summit, and is surrounded on three sides by vaguely columnar rock, slightly suggestive of rude walls. "In regard to the name of the Yaksha, viz., Vakula, which in modern Hindi becomes Bakula, it is remarkable to find the local survival of this name and the awe in which it is still held. Immediately behind Uren is the mouth of a pass which leads into the wild Singhol Hills; and the pass and the hills beyond were the retreat of banditti till long after the Muhammadan invasion. The older banditti are popularly alleged by the villagers to have been cannibals, and their raids are still spoken of by the lowlanders here with dread. These highland aborigines were formerly called *rakshas* or ' demons ' by the plains-people, and the oldest settlement of these *raksha* or *yaksha* tribes is about five miles beyond the mouth of the pass, and is called Bakura—which is identical with the name of the ' *yaksha* ' given by Hiuen Tsiang—*l* and *r* being interchangeable. It is a common practice to name villages after their founders: thus, Bakura village—' the village of Bakura '. And so great was the dread inspired by this Bakura, that he is even now worshipped by the semi-aborigines of the plains (the Dosadhs and Goalas) at a shrine in the village of Jalalabad, about eight miles east from Uren, under the name of Ban-Bakura Nath or the ' Savage Lord Bakura '. His image is in basalt and represents a squat muscular man in a semi-sitting posture. He has a large sensual head, thick lips and curly hair, which latter is fastened in a coil with a scimitar-shaped dagger, as with the aborigines in the Bharhut sculptures. "

To the north of this spot is a footprint in the rock and five yards above it a mound of bricks which may mark the remains of a small stupa as described by Hiuen Tsiang. The impression made on the rock by Buddha's *lota* and the footprints of the Yaksha, which were known to the villagers as Lorik's *lota* mark and footprints, have been destroyed by blasting and the colossal statue of Buddha has also disappeared; but in the spot mentioned by Hiuen Tsiang, Buddha's promenade can still be seen, viz., a narrow level tract between two long massive shoulders of rock. " Before the great accumulation of debris had taken place, the rocks on either side must have stood up like walls and bounded a rocky lane—a most suitable promenade for the great ascetic, affording an outlook only to the distant hills and overhead the sky. "

Further details will be found in Colonel Waddell's article *Discovery of Buddhist Remains at Mount Uren in Mungir (Monghyr) District* (published in the Journal of the Asiatic

Society of Bengal, Part I, 1892) ; and it will be sufficient to say that he rests the identification of the Uren Hill with the hill described by Hiuen Tsiang on the geographical position and physical conformation of the hill; the actual presence and co-existence of all the numerous and specialized remains and rock-markings noted by Hiuen Tsiang; the very numerous votive Buddhist statues and chaityas, and the thousands of names carved on rock, indicating a sacred place of Buddhist pilgrimage; the survival of the old tradition recorded by Hiuen Tsiang that the hill-top was the abode of a demon, the fact that his abode and footprints and the *lota* mark are still pointed out; and the survival of the name and worship of " the savage Lord Bakura ".

Running out from the northern base of the Uren Hills is a small flat and somewhat rocky spur, on the northern extremity of which is situated the village of Uren. Occupying the north-eastern portion of this spur and adjoining the base of the hill, is a terraced area of broken bricks, fragments of Buddhist statues and hewn stones, locally known as the fort of Indradaun. Indradaun, or Indradyumna, was the reigning king of Magadha at the time of the Muhammadan invasion, and he is believed to have been one of the Pala dynasty, which was Buddhist. The whole appearance of the place seems to justify the belief that the so-called *garh* or fort was originally a Buddhist monastery. It seems to have been an almost solid mass of brick buildings, and it contains numerous fragments of Buddhist statues and rough-hewn lintels and door- jambs.

The following description of the remains still extant is given by Dr. Bloch, Superintendent, Archæological Survey, Eastern Circle : " They consist principally in tracings of stupas and other religious marks or emblems, which are found in a great number all over a small granite hill to the south-east of the village. The design of those stupas agrees so closely with the well-known type of the later period of Buddhist architecture, that they cannot be much anterior to the time of Hiuen Tsiang's visit, if they are not even later. A great many of the tracings probably have been destroyed by stone-cutters, who still seem to use this small hill as a quarry ; others have become very indistinct, as also has been the fate of a few rock inscriptions. A collection of statues is found a little to the west of the hill, close to the village. They were all Buddhist, with the exception only of one Gauri-Shankar and one Agni. I did not see many of the statues and inscriptions mentioned by Dr. Waddell. Probably the villagers have taken them away to their houses, where, I was told, a great many ancient statues are now kept and worshipped. Excavation here, as suggested by Dr. Waddell, probably would meet with great opposition on the part of the villagers, and with very little result. The largest mound, and probably the site of various monasteries and temples, is now covered by the modern village of Uren."*

* Report Arch. Surv. Bengal Circle, 1902-3.

APPENDICES.

I.—List of Railway stations.

II.—List of post offices.

III.—List of telegraph offices.

IV.—Principal roads and bungalows.

APPENDIX I.

List of Railway Stations.

(i) *East Indian Railway.*

Main line.

Station.	Distance from Calcutta.
1	2
	Miles.
Simultala ...	217
Jhajha	228
Gidhaur ...	235
Jamui	244
Mananpur ...	253
Kiul Junction ...	262
Lakhisarai ...	262
Mankatha ...	266
Burhee	272

Loop line.

Station.	Distance from Calcutta.
1	2
	Miles.
Bariarpur ...	291
Jamalpur ...	297
Dharhara ...	304
Abhaipur ...	311
Kajra ...	315
Kiul Junction ...	325

South Bihar Railway.

Station.	Distance from Gaya.
1	2
	Miles.
Kiul Junction ...	80
Lakhisarai ...	80
Sirari	70
Sheikhpura ...	64

Monghyr Branch line.

Station.	Distance from Jamalpur.
1	2
	Miles.
Purabsarai ...	4
Monghyr ...	6

APPENDIX I—*contd.*

(ii) *Bengal and North-Western Railway.*

Main line (Cawnpore to Katihar).	
Station.	Distance from Katihar.
1	2
	Miles.
Pasraba	58
Mahesh Khunt	65
Mansi	72
Khagaria	77
Sahebpur Kamal	85
Lakhminia	91
Lakho	98
Begusarai	102
Tilrath	107
Barauni Junction	112
Barauni	113
Teghra	116
Bachhwara	122

Mansi-Bhaptiahi branch.	
Station.	Distance from Bhaptiahi.
1	2
	Miles.
Mansi Junction	61
Padla Ghat	57
Dhamara Ghat	53
Koparia	49
Makhna Bazar	45

Mansi-Samastipur branch.	
Station.	Distance from Samastipur.
1	2
	Miles.
Mansi	60
Khagaria	53
Olapur	47
Imli	42
Salauna	38

Branch lines :—

Sahebpur Kamal-Monghyr Ghat 4m.

Barauni Junction—Mokameh Ghat (Semaria) 5m.

APPENDIX II.

List of Post Offices.

	Asarganj.	
	Bachwara.	
	Baha Chowki ...	C. F. G. K.
	Bakhri Bazar ...	G. K.
	Bamdah ...	C. F. G. K.
T	Barauni.	
T	Barbigha.	
T	Barhiya.	
	Bariarpur.	
T	Basdeopur.	
T	Begusarai.	
	Chakai.	
	Chawara.	
	Cheria-Bariarpur.	
	Chupraon.	
	Dharahara.	
T	Gidhaur.	
	Gogri.	
	Haveli-Kharagpur.	
	Husaina ...	C. G. K.
	Husainabad ...	G. K.
T	Jamalpur.	
T	Jamui.	
T	Jhajha.	
	Kajra.	
T	Khagaria.	
	Khaira.	
	Khizirchak ...	C. F, G. K.
	Kutchery Shampur	C. F. G. K.
T	Lakhisarai.	
T	Lakhminia.	
	Mahdowli ...	K.
	Maheshkhunt ...	K.
	Mallehpur ...	K.
	Mananpur ...	B. C. E. F. G. K.
	Manjhaul.	
	Mankatha ...	B. E. F. G. K.
	Mansi.	
	Mansurchak ...	G.
T	Monghyr.	
	Monghyr Town.	
	Mushkipur ...	G. K.
	Nawakothi ...	F. G. K.
	Nimi ...	K.
	Nurpur ...	C. G. K.
	Pachna ...	G. K.
	Parihara.	
	Pipra Kutchery...	C. F. G. K.
	Rampur Motihari	C. F. G.
	Sahebpur Kamal.	
	Semaria Ghat.	
	Shambhu.	
T	Sheikhpura.	
T	Sheikhpura City.	
	Sikandra.	
	Simri-Bakhtiarpur	K.
T	Simultala.	
T	Surajgarha.	
	Tarapur	K.
T	Teghra.	
	Ulao	

T Combined Post and Telegraph Office (*see* Appendix III.).
B No insurance from or to.
C No insurance from.
E Does not pay money orders.
F Does no savings bank work.
G Does not accept foreign parcels for despatch.
K Does not accept telegrams.

APPENDIX III.

List of Telegraph Offices.

Telegraph Office		Hours of business.			
		Ordinary week-days.		Sundays and Public Holidays.	
Abhaipur R. S.	...	6—21		6—21	
Bachhwara R. S.	...	0—24		0—24	
Balda Ghat R. S.	...	7—21		7—21	
Barauni P. O.	...	7—9	12—17	8-30—9-30	16-30—17-30
Do. R. S.	...	0—24		0—24	
Barbigha P. O.	...	7—9	12—17	8-30—9-30	16-30—17-30
Barhiya P. O. (Burhee)		7—10	13—17	8-30—9-30	16-30—17-30
Basdeopur P. O.	...	7—10	13—17	Closed.	
Begusarai P. O.	...	7—9	12—17	8-30—9-30	16-30—17-30
Do. R. S.	...	7—21		7—21	
Burhee R. S.	...	6—21		6—21	
Dhamara Ghat R. S.		7—21		7—21	
Dharahara R. S.	...	6—21		6—21	
Dudhaila R. S..	...	7—21		7—21	
Dumra R. S.	...	6—21		6—21	
Gidhaur P. O.	...	7—9	12—17	8-30—9-30	16-30—17-30
Do. R. S.	...	6—21		6—21	
Imli R. S.	...	7—21		7—21	
Jamalpur P. O.	...	8—18		8—10	16—18
Do. R. S.	...	6—22		6—22	
Jamui R. S.	...	6—21		6—21	
Do. P. O.	...	7—9	12—17	8-30—9-30	16-30—17-30
Jhajha P. O.	...	7—10	13—17	8-30—9-30	16-30—17-30

APPENDIX III—*concld.*

Telegraph Office.	Hours of business.	
	Ordinary week-days.	Sundays and Public Holidays.
Jhajha R. S. ...	6—21	6—21
Kajra R. S. ..	6—21	6—21
Khagaria P. O. ...	7—9 12—17	8-30—9-30 16-30—17-30
Kiul R. S. ...	6—22	6—22
Koparia R. S. ...	7—21	7—21
Lakhisarai P. O. ...	7—9 12—17	8-30—9-30 16-30—17-30
Do. R. S. ..	6—21	6—21
Lakhminia P. O. ...	7—9 13—17	8-30—9-30 16-30—17-30
Do. R. S. ...	7—21	7—21
Mahesh Khunt R. S.	7—21	7—21
Makhana Bazar R. S.	7—21	7—21
Mananpur R. S. ...	6—21	6—21
Mankatha R. S. ...	6—21	6—21
Mansi R. S. ...	7—21	7—21
Monghyr P. O. ..	7—20	8—10 16—18
Do. R. S. ...	6—21	6—21
Do Ghat R. S.	7—21	7—21
Olapur R. S. ...	7—21	7—21
Pasraha R. S. ...	7—21	7—21
Sahebpur Kamal R. S.	7—21	7—21
Semaria Ghat R. S. ..	0—24	0—24
Sheikhpura P. O. ...	7—9 12—17	8-30—9-30 16-30—17-30
Do. P. O. ...	6—21	6—21
Simultala P. O. ...	7—9 12—17	8-30—9-30 6-30—17-30
Do. R. S. ...	6—21	6—21
Sirari R. S. ...	6—21	6—21
Teghra P. O. ...	7—9 13—17	8-30—9-30 16-30—17-30
Do. R. S. ...	7—21	7—21
Tilrath R. S. ...	7—21	7—21

APPENDIX IV.

PRINCIPAL ROADS AND BUNGALOWS.

Road.	Length.	Class.	Bungalows.
1.	2	3	4
	M. F.		
1. Ghoreghat-Burhee (the Bhagalpur-Patna road).	18 5	IA	Bariarpur.
	35 5	IIB	Monghyr.
			Baha.
			Surajgarha.
			Lakhisarai.
			Burhee.
2. Bariarpur (10 miles from Monghyr on road no. 1 above) to Jamui.	39 5	IA	Kharagpur.
			Gangta.
			Gurdih.
			Mallehpur.
			Jamui.
3. Sultanganj* to Belhar ...	11 4*	IA	Tarapur.
	11 4	IIB	Sangrampur.
4. Lakhisarai to Arha ...	2 5	IA	Lakhisarai.
	13 7	IIA	Sikandra.
	13 5	IIB	
5. Lakhisarai to Jamui ...	18 0	IV	Lakhisarai.
			Majhwe.
			Jamui.
6. Ramgarh (6th mile of road no. 4 above) to Sheikhpura.	14 0	IIA	

* The first three miles of this road from Sultanganj are in Bhagalpur district.

APPENDIX IV—*contd.*

Road.	Length.	Class.	Bungalows.
1	2	3	4
	M. F.		
7. Sheikhpura-Barbigha ...	11 0	IA	Barbigha.
8. Sheikhpura-Jianbigha (Giriak road).	9 0	IIA	
9. Jamui-Chakai ...	31 4	IIB	Tehia.
			Betya.
			Chakai.
10. Jamui-Nawadih ...	4	IA	
	8 0	IIB	
11. Jamui-Sheikhpura ...	5 7	IA	Girinda.
	29 0	IIA	Sikandra.
			Chewara.
12. Jamui-Harkhar ...	7 0	IIA	Bissunpur.
	13 0	IV	
13. Simultala-Chakai ...	14 6	IIA	Kankuria.
	1 4	III	
14. Monghyr Ghat-Rasidpura (Tirhut road).	14 6	IA	Ballia.
	28 2	IIA	Begusarai.
			Teghra.
			Bachhwara.
15. Monghyr Ghat-Khagaria	7 0	IIB	
	2 0	IV	
16. Khagaria-Shakarpura ...	19 0	IIB	Parihara.
			Bakhri.
17. Mansi-Balha (Nepal road)	24 6	IIB	Dhamara.
			Bakhtiarpur.
18. Koperia-Sonbarsa ...	6 4	IIB	

APPENDIX IV—*concld.*

Road.	Length.	Class.	Bungalows.
1	2	3	4
	M. F.		
19. Gogri-Bhatkar ...	2 3	IA	Gogri Jalpur.
	3 5	IIA	
	8 4	IV	
20. Begusarai-Manjhaul ...	10 3	IB	
21. Begusarai-Shambho ...	8 4	IIB	Shamho.
22. Manjhaul-Garhpura ...	8 0	IIB	Garhpura.

Class IA—Metalled roads, bridged and drained throughout.

IB—Metalled, partially bridged and drained.

IIA—Unmetalled, bridged and drained throughout.

IIB—Unmetalled, partially bridged and drained.

III—Roads banked and surfaced with gravel, but not drained.

IV—Roads banked but not surfaced, partially bridged and drained.

INDEX.

C

D

J

K

N

O

P

Q

R